# DECENTRALIZED PULSE SIGNAL DECOMPOSED SYNCHRONIZER FOR INVERTER MAINTENANCED GRID POWER

PARESH RAWAL

# Contents

# List of Figures

# List of Tables

# Abbreviations

| | | |
|---|---|---|
| RE | - | Renewable Energy |
| IRENA | - | International Renewable Energy Agency |
| VAR | - | Volt Ampere Reactive |
| IDM | - | Inverter Dominated Microgrid |
| IEEE | - | Institute of Electrical and Electronics Engineers |
| *CIGRÉ* | - | International Council on Large Electric Systems |
| AC | - | Alternating Current |
| DC | - | Direct Current |
| EV | - | Electric Vehicle |
| G2V | - | Grid to Vehicle |
| V2G | - | Vehicle to Grid |
| ICT | - | Information and Communication Technology |
| PCC | - | Point of Common Coupling |
| PWM | - | Pulse Width Modulation |
| FACTS | - | Flexible Ac Transmission Systems |
| PI | - | Proportional Integral |
| PID | - | Proportional Integral Derivative |
| PR | - | Proportional Resonance |
| $\alpha\beta$-PR | - | Stationary Reference Frame Control |
| DPC | - | Direct Power Control |
| DTC | - | Direct Torque Control |
| MPC | - | Model Predictive Control |
| PLL | - | Phase Locked Loop |
| SRF PLL | - | Synchronous Reference Frame PLL |
| dSRF PLL | - | Double Synchronous Reference Frame PLL |
| DSOGI PLL | - | Dual Second Order Generalized Integrator PLL |
| EPLL | - | Enhanced PLL |
| MAF PLL | - | Moving Average Filter Based PLL |
| MCS | - | Multi-Component Signals |
| FT | - | Fourier Transform |
| FFT | - | Fast Fourier Transform |
| DFT | - | Discrete Fourier Transform |
| EMD | - | Empirical Mode Decomposition |
| WT | - | Wavelet Transform |
| EWT | - | Empirical Wavelet Transform |

| | | |
|---|---|---|
| WPT | - | Wavelet Packet Transform |
| VMD | - | Variational Mode Decomposition |
| IMF | - | Intrinsic Mode Functions |
| HHT | - | Hilbert Huang Transform |
| STFT | - | Short Time Fourier Transform |
| ST | - | Stockwell's Transform |
| LED | - | Light Emitting Diode |
| DSO | - | Digital Storage Oscilloscope |
| *RMSE* | - | Root Mean Square Error |
| αβPLL | - | Stationary Reference Frame PLL |
| SOGI PLL | - | Second Order Generalized Integrator PLL |
| ddSRF PLL | - | Decoupled Double Synchronous Reference Frame |
| dαβPLL | - | Double Stationary Reference Frame PLL |
| FPDdαβPLL | - | Fractional Phase Divider Double Stationary Reference Frame PLL |
| Advanced FDPdαβPLL | - | Advanced Fractional Phase Divider Double Stationary Reference Frame PLL |
| EMAFPLL | - | Enhanced Moving Average Filter PLL |
| EMAFPLL | - | Enhanced Moving Average Filter PLL |
| MSHDCPLL | - | Multi Sequences Harmonic Decoupling Network Based PLL |
| DNαβPLL | - | Decoupling Network Stationary Reference Frame PLL |
| EPMAFPLL | - | Enhanced Phase Error Estimation Moving Average Filter PLL |
| αβEPMAFPLL | - | Stationary Reference Frame Enhanced Phase Error Estimation Moving Average Filter PLL |
| BMPC | - | Bidirectional Model Predictive Control |
| GTC | - | Grid-Tied Control |
| GFC | - | Grid-Forming Control |
| HIL | - | Hardware-in-Loop |
| DA | - | Dynamic Adaptability |
| DAMPC | - | Dynamically Adaptable Model Predictive Control |

# List of Symbols

| | | |
|---|---|---|
| $f^*$ | - | Grid frequency |
| $\theta^*$ | - | Grid voltage phase |
| $i^*$ | - | Reference current |
| $v_{ref}$ | - | Reference voltage |
| $i$ | - | Actual Current |
| $f$ | - | MCS input signal |
| $K$ | - | Number of modes |
| $u_k$ | - | Modes |
| $f_K$ | - | Fourier Transform of $f$ |
| $\omega_k$ | - | Central frequency of mode |
| $\Delta\omega_k$ | - | Band width of a mode |
| $u_k^A$ | - | Analytic function from Hilbert Transform of $u_k$ |
| $u_k^D$ | - | Frequency-translated form of $u_k^A(t)$ |
| $\alpha$ | - | Balancing parameter of the data-fidelity constraint |
| $\mu$ | - | Lagrangian multiplier. |
| $n$ | - | Iteration step |
| $\tau$ | - | Tolerance of convergence |
| $v_{(ph)x}$ | - | Input voltage signal (where $x$ indicates the test case serial number) |
| $u_1$ | - | Extracted first mode (fundamental frequency component) |
| $\varepsilon_v$ | - | Absolute error in magnitude |
| $\varepsilon_\theta$ | - | Absolute error in phase |
| $\varepsilon_f$ | - | Absolute error in frequency |
| $v^*_{rms}$ | - | RMS value of nominal grid voltage |
| $v_{rms}$ | - | RMS value of extracted fundamental voltage |
| $\theta^*$ | - | Phase angle of nominal grid voltage |
| $\theta^*$ | - | Phase angle of extracted fundamental voltage |
| $F^*$ | - | Frequency of nominal grid voltage |
| $F$ | - | Frequency of extracted fundamental voltage |
| $\sigma^*$ | - | Standard deviations of the sampled values of the nominal grid voltage |
| $\sigma$ | - | Standard deviations of the sampled values of the extracted fundamental frequency component |
| $f_{ist}$ | - | Instantaneous frequency |
| $\Delta_{trans}$ | - | Frequency transition time |

| | | |
|---|---|---|
| $f(t)$ | - | Sampled input to EWT synchronizer |
| $F(\omega)$ | - | Frequency spectrum of $f(t)$ |
| $\omega_n$ | - | Central frequency of nth mode of EWT synchronizer |
| $\Lambda_i$ | - | $i^{th}$ Segment of frequency spectrum |
| $\omega_{i,}$ | - | Mid frequency |
| $T_n$ | - | Transition Phase |
| $\beta(y)$ | - | *Mayer's* auxiliary function |
| $\hat{\emptyset}_n(\omega)$ | - | *Mayer's* mother wavelet function |
| $\hat{\varepsilon}_n(\omega)$ | - | *Mayer's* scaling wavelet function |
| $u_{dc}$ | - | DC Voltage |
| $u_{inv}$ | - | Inverter pole voltage |
| $i_x$ | - | Per phase inverter current |
| $u_x$ | - | Per phase grid voltage |
| $L_{fx}$ | - | Per phase equivalent filter inductance |
| $R_{fx}$ | - | Per phase equivalent filter resistance |
| $x$ | - | Phase sequence |
| $a$ | - | $e^{j2\pi/3}$ |
| $s_{a,}$ | - | Status of the top switch of the inverter leg a |
| $s_b$ | | Status of the top switch of the inverter leg b |
| $s_c$ | - | Status of the top switch of the inverter leg c |
| $u_z$ | - | Inverter voltage vectors |
| $F_s$ | - | Sampling frequency |
| $T_s$ | - | Sampling time |
| $K$ | - | Sampling instance |
| $u_{inv}(k)$ | - | Inverter leg voltage at $k^{th}$ sample |
| $i_x(k)$ | - | Per phase inverter current at $k^{th}$ sample |
| $u_x(k)$ | - | Per phase grid voltage at $k^{th}$ sample |
| $i^p(k+1)$ | - | Current predictions for $k+1^{th}$ sample |
| $G$ | - | Weight function |
| $f_m$ | - | Primary target of weight function |
| $f_n$ | - | Secondary target of weight function |
| $\lambda$ | - | Weighting parameter for secondary target |
| $i^*{}_\alpha(k+1)$ | - | $\alpha$ component of reference current, $i_x{}^*$, in stationary reference frame for the $k+1^{th}$ sampling instance |
| $i^*{}_\beta(k+1)$ | - | $\beta$ component of reference current, $i_x{}^*$, in stationary reference frame for the $k+1^{th}$ sampling instance |
| $i_\alpha(k+1)$ | - | $\alpha$ component of actual current, $i_x$, in stationary reference frame for the $k+1^{th}$ sampling instance |

| | | |
|---|---|---|
| $i_\beta(k+1)$ | - | β component of actual current, $i_x$, in stationary reference frame for the $k+1^{th}$ sampling instance |
| $i^p{}_\alpha(k+1)$ | - | α component of Predicted current, $i_x{}^p$, in stationary reference frame for the $k+1^{th}$ sampling instance |
| $i^p{}_\beta(k+1)$ | - | β component of Predicted current, $i_x{}^p$, in stationary reference frame for the $k+1^{th}$ sampling instance |
| $i^*{}_\alpha$ | - | α component of reference current, $i_x{}^*$, in stationary reference frame |
| $i^*{}_\beta$ | - | β component of reference current, $i_x{}^*$, in stationary reference frame |
| $i_\alpha$ | - | α component of actual current, $i_x$, in stationary reference frame |
| $i_\beta$ | - | β component of actual current, $i_x$, in stationary reference frame |
| $u_\alpha$ | - | α component of grid voltage, $u_x$, in stationary reference frame |
| $u_\beta$ | - | β component of grid voltage, $u_x$, in stationary reference frame |
| $u_\alpha{}^*$ | - | α component of reference voltage in stationary reference frame |
| $u_\beta{}^*$ | - | β component of reference voltage in stationary reference frame |
| $u_\alpha(k+1)$ | - | α component of grid voltage, $u_x$, in stationary reference frame for the $k+1^{th}$ sampling instance |
| $u_\beta(k+1)$ | - | β component of grid voltage, $u_x$, in stationary reference frame for the $k+1^{th}$ sampling instance |
| $F_{asw}$ | - | Average sampling frequency |
| $mag_i$ | - | Magnitude tracking accuracy |
| $ang_i$ | - | Angle tracking accuracy |
| $THD_i$ | - | Total Harmonic Distortion of current |
| $P_{loss}$ | - | Inverter Power Loss |
| $i^*{}_{xrms}$ | - | rms value of the reference current |
| $i_{xrms}$ | - | rms value of the actual current |
| $\theta^*$ | - | Phase angles of the reference current |
| $\theta$ | - | Phase angles of the actual current |
| $i_{err}$ | - | Instantaneous current error |
| $i_{aerr}$ | - | Instantaneous current error for phase a |
| $i_{dc}$ | - | DC current |
| $C_{dc}$ | - | DC link capacitance |
| $P^*$ | - | Active power reference |
| $P$ | - | Active power |
| $Q^*$ | - | Reactive power reference |

| | | |
|---|---|---|
| $Q$ | - | Reactive power |
| $\omega$ | - | Line frequency |
| $\delta$ | - | Phase angle between $u_x$ and $u_{inv}$. |
| $u_L$ | - | Inductor voltage drop |
| $u_x(k-1)$ | - | Grid voltage, $u_x$, for k-1$^{th}$ sample |
| $p_{\alpha\beta}$ | - | Stationary reference frame coordinates of active power |
| $q_{\alpha\beta}$ | - | Stationary reference frame coordinates of reactive power |
| $SoC$ | - | State of charge of battery |
| $C$ | - | C rating of battery |
| $t_d$ | - | Duration of charge/discharge operation, |
| $fast/slow$ | - | Rate of charging |
| $I_{dmax}$ | - | Maximum value of battery current |
| $m$ | - | Multiplication factor suggested my battery manufacturer |
| $L_{eq}$ | - | Per phase equivalent inductance |
| $R_{eq}$ | - | Per phase equivalent resistance |
| $L_{gx}$ | - | Per phase equivalent grid inductance |
| $R_{gx}$ | - | Per phase equivalent grid resistance |
| $g$ | - | Grid side parameter |
| $f$ | - | Filter parameter |
| $s_x^p(k+1)$ | - | Status of the top switches of the inverter legs predicted for $k+1^{th}$ instant |
| $s_x(k)$ | - | Status of the top switches of the inverter legs for $k^{th}$ instant |
| $Genmix$ | - | Generation mix of IDM |
| $LOH$ | - | Lowest order harmonics |
| $\varepsilon_u$ | - | RMS voltage error |
| $THD_u$ | - | Voltage harmonics |
| $\rho_H$ | - | Priority signals to meet harmonic standards |
| $\rho_L$ | - | Priority signals for inverter losses |
| $\rho_T$ | - | Priority signals for reference tracking |
| $i^*_{a-gf}$ | - | Reference current in grid-forming mode for phase a |

# Abstract

High penetration of power electronics in power generation is envisaged to revamp the legacy grid to Inverter Dominated Microgrids (IDM). Such a migration bringsforth unique challenges in grid synchronization and inverter power control arenas. An investigation on Variational Mode Decomposition (VMD) technique extended to grid synchronization of power converters in emerging non-linear grids is done. Number of modes and data fidelity parameter of VMD are tailored to achieve finer separation of the fundamental frequency in spite of the spectral band variations and minor frequency deviations of the voltage. The test cases are formulated based on CIGRÈ and IEEE Task Force 1159.2 repository on the foreseen power quality issues in the emerging power utilities. Two statistical indices, absolute error in percentage and RMS error, are chosen to assess the compliance of extracted signals with the applied values at cycle as well as sample level. A significant reduction in the tracking time in all the test conditions has been observed in simulation when compared to the conventional grid synchronizers. Appreciable accuracy levels are obtained in tracking the required signal attributes of the input signal. The frequency decomposition approach of VMD synchronizer exhibited an immunity towards random transient events like voltage surges, zero crossing disorders, phase and frequency jump, etc. Further, validation in hardware has been conducted with emulated grid signals representing a wide range of transient events and the results are found satisfactory.

Next, Model Predictive Control (MPC) has been identified as an upcoming converter control that can accomplish additional capabilities such as multi-mode operation, self-adaptability, high efficiency, seamless bi-directional power transfer, etc. besides the prime objective of reference tracking. But, implementing MPC in grid tied converter control demands its design parameters like sampling frequency, weight function coefficients and model parameters to be aptly chosen to match the hardware intricacies. So, a primary investigation has been carried out to identify the correlation between sampling rates of MPC for desired performance indices like tracking accuracy, average inverter switching frequency, current harmonic profile and inverter power loss applied to a three phase grid connected inverter. It is observed that the tracking accuracy greatly depends on the prediction interval, and in turn on the sampling rates. The dynamic performance of the inverter is tested for step changes in reference and source specifications besides conducting the steady state analysis. Extensive analysis for various combinations of MPC design parameters is presented. The results exhibited high reference tracking accuracy with improved harmonic profile for high to moderate sampling frequencies, but with a tax on the inverter efficiency. The choice of switching frequency for a trade-off between high tracking accuracy and inverter efficiency has also been suggested.

Next, a bidirectional MPC with improved dynamic response has been developed. The concept of instantaneous power theory is merged with the MPC to track a power reference even without any power loop in the controller. A reference current, synthesized from the power references to serve as the forcing quantity in the inner loop, contributed to enhanced dynamic response of the bidirectional MPC when migrating from inverter to converter mode and vice versa. The formulated control has been applied for control of three phase bidirectional converter in a grid tied solar PV system as well as for charge/discharge control in a battery energy storage application. The effectiveness of the control for steady state and dynamic conditions have been evaluated in terms of tracking accuracy, harmonic profile, converter power loss and transition time. The test results demonstrated extremely small transition time with accurate tracking, better harmonic profile and reduced converter losses.

Further, design and development of *Dynamic Adaptability* feature that enables multi-mode operation and self-adaptability to model predictive controlled grid-tied inverters in IDM environment has been carried out. The dynamically adaptable MPC has a multi-objective weight function targeting switching frequency reduction along with reference tracking. The secondary function weight has been dynamically varied in accordance with the demand of the operating condition for self-reliant change-over across grid-tied and grid-forming modes in IDM environment. A customized dynamic adaptability database has been developed through prior investigation on the IDM, considering possible operating conditions with diverse combinations of the sampling frequency and the secondary function weight. Investigations unveiled that there is a unique value of secondary weight for every operating condition, beyond which the reference tracking is lost. This necessitated the dynamic assignment of the weight to achieve a desirable inverter performance. The dynamic weightage assignment has been accomplished by accessing the Dynamic Adaptability (DA) database through the priority signals separately for harmonic standards, inverter losses and reference tracking. The DA imparted MPC is tested for its mode transfer capability with desired performance under all operating modes of IDM in simulations with a 15 kVA grid-tied/forming inverter and the results are presented. Further, the experimental investigations have been conducted with a 10 kVA inverter on a 5-bus laboratory scale IDM emulator along with hardware-in-loop. Both inter-mode and intra-mode transitions were initiated in the IDM and the inverter performance has been studied. Both the simulation and the hardware results yielded competent values of reference tracking accuracy and mode transition times.

# Chapter 1

# Introduction

## 1.1 Introduction

Renewable energy (RE) technology emerged in the last decade as an ideal choice for electrical energy production owing to its omnipresence with no carbon footprint. The global installed capacity of the renewable energy fed power system has seen an escalation of 20% in the first two decades of the twenty first century [1]. IRENA's renewable energy projection for 2050 shown in Figure 1.1 reckons the possibility of RE contribution moving towards 100% [2].

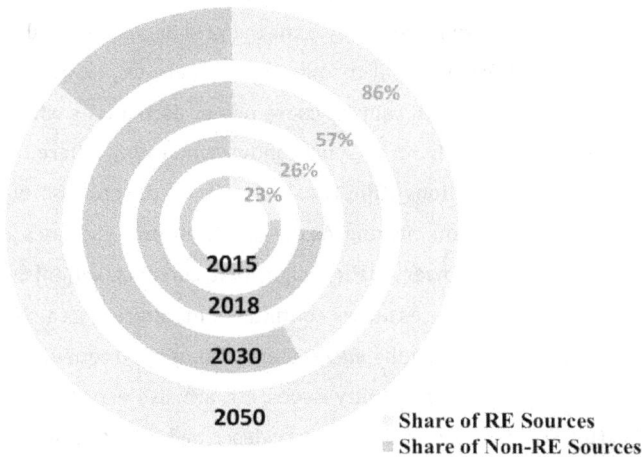

Figure 1.1: RE 2050 roadmap by IRENA [2].

Under such condition, the instantaneous RE power penetration has to be significantly large in order to achieve a significant RE contribution in terms of percentage energy penetration to the grid [2], [3]. This is mainly due to the diurnal cycles of solar radiation and seasonal cycles of wind. This may lead to a condition where the instantaneous renewable power injection may exceed the synchronous power during peak RE generation [4]. Besides, nearly all renewable energy sources including solar, wind, etc. are interfaced to the grid through power processor/s like inverters leading to a large power electronic invasion into the power grid [5]. Large scale diffusion of power electronic devices reduces the inertial response of the grid and can cause frequency excursions that result in cascaded generator outage and blackout.

1

In 2018, almost all countries recast their existing renewable energy policies and plans in support of greater fleets of renewable energy capacity addition [3], [6], [7]. Iceland and Paraguay have already accomplished the transition to 100% RE grid while Costa Rica and Norway nearing the finishing line [1]. As per IRENA's survey yet other 140 countries are in the path towards achieving this goal by 2050 [8].

Such a paradigm shift brings about a new set of configuration, control, and operational challenges to both the power grid and the RE interfacing inverter. This thesis investigates some of these challenges.

## 1.2   Impact of high RE penetration on legacy grids

Legacy grid had been a simple, linear system with definite voltage and frequency profile. The RE sources were integrated synchronously onto the grid as a sinusoidal current source that followed the voltage and frequency at its output terminals, i.e., as grid following units. The intermittency and variability of RE sources cause power oscillations which can challenge the grid stability. However, high share of synchronous generation offered high inertia which leveraged the low RE penetrations [9].  Thus, a low-level penetration of RE onto the grids required no specific control or monitoring. As the RE penetration escalates, issues such as large generation uncertainty, low power quality, high fault currents, diminished inertia etc. will challenge the grid stability. It necessitates continuous monitoring and sophisticated control. Additional challenges like voltage imbalances, thermal shut off, frequency deviations, low fault tolerance, unintended islanding, reliability issues, etc. are also expected to arise with high RE penetration [4], [10]. Some of these challenges are described in the following section.

### 1.2.1 Key issues and challenges with high RE penetrations

Issues and challenges due to high RE penetration are as listed:

1.   Reduction in inertia
2.   Coordinated control of power electronics at high penetration
3.   Frequency fluctuations
4.   Voltage distortions
5.   New age power quality issues
6.   Intermittency and variability of RE
7.   Unintentional disconnections

The large kinetic energy of the turbo generators counteracts any change or disturbance in the system and provides transient support to maintain the grid stability and to enhance reliability [11]. But as the grid evolves with large RE share through large number of power electronic devices, the inherent inertial response will decline. This makes the system more sensitive to even small disturbances, thus challenging the transient and dynamic grid stability. Poor dispatchability of RE sources added to the low inertia will cause the grid frequency to fluctuate extensively [12]. Supplying non-linear loads with large VAR demand is yet another challenge as power converters as RE interface mostly inject active power with unity power factor. These may not be capable of compensating the huge VAR demand and will cause asymmetric distortions in line voltage like sag/swell/unbalance at distribution points [13].

Arbitrary harmonic composition may arise due to multiple switching schemes, different frequencies, inadequate filters, etc. [5]. Thus, uncertain harmonic profile with new age power quality issues like interharmonics, subharmonics, supraharmonics, frequency and phase jump, random transients, multiple zero crossings, etc. are expected with high RE penetration [14]. These impose additional challenges in power converter control schemes and grid synchronization of power converters which may lead to inefficient power system operation and control. Variability and intermittency of RE sources makes the *load following* difficult. This can cause a persisting gap between generation and demand which leads to unplanned outages, non-reliable operation and even up to grid collapse. This necessitates large spinning reserves or fast responding grid connected energy storage systems to be supplemented in the grid.

Search for the solutions to most of these problems has taken the research towards the possibilities of microgrid; an innovative concept that involves dispatchable generation/load on the grid.

## 1.3 RE development through microgrids

Microgrids are considered as a viable solution which facilitates high penetration of RE without burdening the legacy grid operations. Consequently, these co-exist with the legacy power grids and share the escalating capacity addition required. Often these microgrids are proliferated with large number of power converters and are hence referred to as Inverter Dominated Microgrids (IDM) [15]. The upcoming sections describe the fundamental concepts, components, issues, and research problems in IDMs.

## 1.3.1 Microgrids

Microgrids can be described by topologies, functions, type of power supply, locations, size, operating modes, etc. Hence a variety of definitions have been propounded by different international agencies, some of which are given below:

*(i). IEEE Standards Association: Microgrids are localized grids that can disconnect from the traditional grid to operate autonomously. Because they are able to operate while the main grid is down, microgrid can strengthen grid resilience and help mitigate grid disturbances as well as function as a grid resource for faster system response and recovery* [5].

*(ii). US Department of Energy: A microgrid is a group of interconnected loads and distributed energy resources within clearly defined electrical boundaries that acts as a single controllable entity with respect to the grid. A microgrid can connect and disconnect from the grid to enable it to operate in both grid-connected and islanded mode* [5].

*(iii). Microgrid Institute: A microgrid is a small energy system capable of balancing captive supply and demand resources to maintain stable service within a defined boundary. There's no universally accepted minimum or maximum size for a microgrid* [5].

*(iv). CIGRÉ: A microgrid is an electricity distribution system containing loads and distributed energy resources (such as generators, storage devices or loads) that can be operated in a controlled, coordinated way while islanded from any utility grid, and is under the control of a single management entity* [5].

## 1.3.2 Microgrid classification

Microgrid can be classified based on the type of power supply as AC microgrid, DC microgrid and hybrid microgrid. Microgrids can also be classified based on their mode of operation as grid-tied microgrid, islanded microgrid and hybrid microgrid with transition across grid-tied and islanded mode.

## 1.3.3 Microgrid components

A generalized microgrid network is presented in Figure 1.2. A brief description of this system is given below:

- **Micro-generators:** Microgrid has a heterogeneous generation mix comprising of conventional sources and RE sources. The conventional generators include diesel generators, micro-turbines etc. which are dispatchable sources, ideal for ramp rate control. Varieties of RE generators like solar photovoltaic, wind, biomass, small-hydro, etc. are also included in the generation mix.

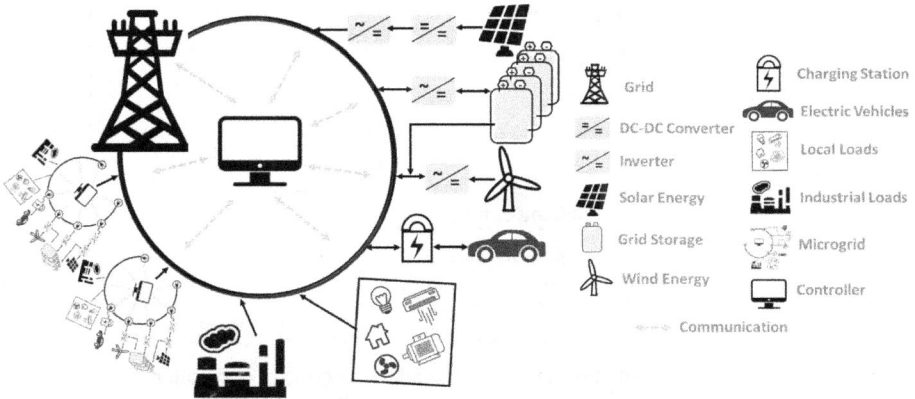

Figure 1.2: Conceptual visualization of a microgrid network

- **Power converters:** Various topologies of power converters like inverters, rectifiers, bidirectional converters, DC-DC converters, etc. are used in a microgrid for power conditioning [14]. These are controlled to perform specific tasks of power conditioning as well as operation and control.

- **Energy storage:** Energy storage is an essential component in any microgrid to compensate the supply-demand gap. Electrochemical and solid-state batteries, pumped hydro, flywheel, super capacitor, compressed air, thermal storage, gravitational storage, etc. are some of the energy storage technologies available [16].

- **Electric Vehicles (EV) charging stations**: The EV batteries, associated power converter and charging stations can be utilized for both Grid-to-Vehicle (G2V) and Vehicle-to-Grid (V2G) applications. G2V helps energy management on microgrids serving as a deferrable load whereas V2G serves occasionally as energy source.

- **Loads:** Residential, commercial and industrial loads in the locality are fed by the microgrid. They can have a linear as well as non-linear voltage and current characteristics. Energy storage during charging mode and EV under G2V mode also appear as loads on microgrid [5].

- **Microgrid central controller:** A central controller and multiple local controllers embedded with multi-tiered supervisory algorithms collectively manage the microgrid [5]. It has to cater to numerous functions like:

  - Generation control

- Frequency control
- Voltage Control
- Controls operating modes of inverter
- Fault ride through control
- Black start control
- Forecasting and scheduling operation of RE sources
- Data collection
- Energy pricing and pricing control
- Energy storage management control
- Demand response

  Research works are still underway to formulate one such system which can unify control requirements and serve as a standard microgrid controller.
- **Information and Communication Technology (ICT):** ICT is a blanket term that describes a network dedicated to carry out multi-way communication among various components. The communication network is categorized as local area network, metropolitan area network or wide area network based on the number of nodes, expanse, data volume and data size. ICT adopts a combination of wired and wireless technologies like Ethernet, fiber optics, power line carrier, Wi-Fi, Bluetooth, ZigBee, Insten, etc. for its data exchange [5].

## 1.3.4 Operating modes of microgrid

Microgrid can operate in grid connected mode or islanded mode [17]–[19]. A brief description of these modes, its significance and challenges are presented below.

- **Grid connected mode:** In grid connected mode, the microgrid is synchronously tied to the main grid and exchange bidirectional power with it. All the distributed generation units in microgrid will follow the voltage and frequency of the mains using grid synchronization techniques. The surplus power in microgrid can be exported to the main grid while the deficit power will be imported from the main grid. The current injected by microgrid in grid connected mode has to meet the synchronization and power quality standards mandated by the grid codes. The local microgrid controller achieves this by employing algorithms with independent active and reactive power control [5], [20].
- **Islanded mode:** When islanded, microgrid works autonomously by forming a local grid to maintaining continuous supply for local loads. Most microgrids in remote locations,

6

ships, military facilities, etc. operate autonomously. Once islanded, local controller decides the operating voltage and frequency of the microgrid network. The disconnection from mains can be either intentional or unintentional [5], [20].

- *Intentional islanding*: This is very much essential for protection of microgrid components. The magnitude of fault currents resulting from even minor grid faults or disturbances may overload the power converters within the *microgrid*. Under such condition a microgrid may choose to disconnect from the main grid and continue supplying the local loads. When islanded, the microgrid inverter initially working in the grid-tied mode needs to alter its control to grid-forming mode. Similarly, upon grid resumption, the inverter control has to revert to grid tied mode. Such inter-mode transitions must adhere to the grid codes specified to the islanding operation [21], [22].

- *Unintentional islanding*: A loss of mains due to load shedding, feeder repair or maintenance, grid fault, etc. while the microgrid is in grid connected mode is termed as unintentional islanding, where the islanding occurs without any prior decision or actuation by the microgrid. The power injected from the microgrid will still energize the feeder and may cause potential rise at Point of Common Coupling (PCC). It can cause harm to the utility workers/ hinder the maintenance work as well. The microgrid controller must monitor the grid status and initiate proper islanding protocols under such unintentional loss of mains [23].

## 1.4. Issues in microgrid operation at high RE penetration

With high penetration of RE sources large dispersion of power electronics will be apparent. A superset of AC microgrid with high penetration of RE and associated fleet of inverters operating in parallel with each other constitute the IDM. IDMs are alternatively called as inverter based power systems or power electronics dominated power systems in numerous recent literatures and technical reports [2], [8], [14], [15], [21]–[28].

Utilizing microgrids for migration towards 100% RE fed utility has gained large momentum in the past decade. A literature review is conducted to understand the postulated future trends in the emergent microgrid [2], [4], [13], [29]–[32]. Vital research problems identified from this study are briefed below:

- With the low inertia and highly dynamic grid structure, ensuring reliability and stability in microgrid is a crucial requirement. Developing integrated controls that improve the stability of microgrid is an important task [9], [12], [33].

7

- Advanced inverter control strategies like reactive power control, frequency support, improved efficiency, voltage support, minor fault ride through, ramp rate control, soft start, etc. [15], [28], [34] other than power control is an essential aspect in high RE penetrated grids.
- Though DC microgrid has no major power quality issues and has low transmission loss, it is currently less desirable as majority of loads are AC. Recent development of DC adaptable loads demands the growth of DC microgrid and its associated control [5], [35]–[37].
- New avenues for energy storage and associated technologies with improved energy density, long storage hours and lower installation cost is another vital research problem to be solved for reliable operation of future microgrid without energy shortage and grid collapse.
- Developing cutting edge energy management systems for controlled operation of energy storage is another research problem being focused by both scientific community and industries [5], [35]–[37].
- Most of the studies pertaining to the socio-economic impact of transition to microgrid has been done on laboratory test beds or small area microgrid. Interoperability costs and overhead expenses, that may arise when deploying a large number of microgrid for supporting the primary energy demand of the world, is still an enigma [5], [35]–[37].
- Incorporation of advanced communication technologies and artificial intelligence features that impart an autonomous decision-making capability to microgrid is expected to bloom exponentially in approaching years [5], [35]–[37].
- Unavailability of uniform standards for integration and interoperability among microgrids is yet another issue.

Amidst these, dominance of inverters anticipated in the state-of-the-art microgrid has been found to be an interesting area of study with contemporary research scope. The traditional inverter control techniques have been reviewed to weigh the merits and demerits to identify the suitability to work in IDMs. Further improvements have been addressed in this research.

*Generic structure of grid-tied inverter control*

Figure 1.3 shows a generic structure of Grid-Tied (GT) inverter system. This system has three main components, viz. the inverter, the power control unit and the grid synchronization unit.

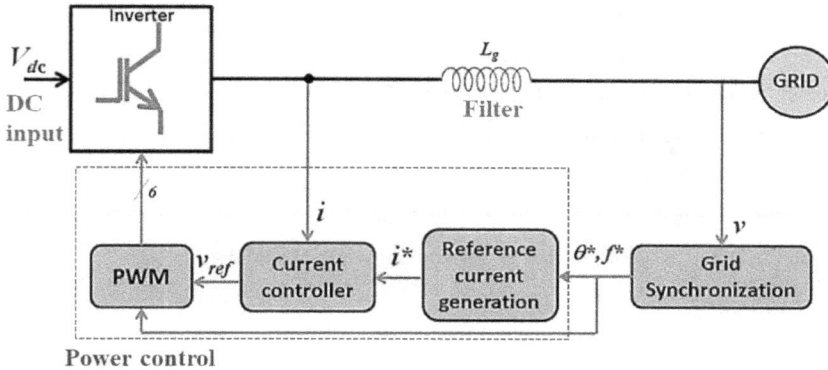

Figure 1.3: Generic structure of GT inverter control [5]

The grid synchronization unit senses the grid voltage, $v$, and determines the grid frequency, $f^*$, and grid voltage phase, $\theta^*$. The power control algorithm governs the inverter's output voltage and frequency, power delivered, power quality, power flow directions etc. Power control has a faster inner current control loop and slower outer reference current computation loop. The magnitude of reference current, $i^*$, is computed from power equations, whereas its shape and phase information are taken from $f^*$ and $\theta^*$. The current controller generates the reference voltage, $v_{ref}$, necessary to reduce the steady state error between $i^*$ and actual current, $i$. PWM block generates the inverter switching patterns using the $v_{ref}$ via various PWM techniques like sine PWM, space vector PWM, etc. [5], [22], [30]. Though the time-tested grid-tied scheme of Figure 1.3 works in the grid-following mode successfully, it is afraid to fall short in meeting some of the advanced control requirements when working in IDMs.

## 1.4.1 Research gaps on control tasks in IDM

In legacy grids, the inverters are extensively used in various grid edge applications like renewable energy conversions, harmonic compensation, uninterrupted power supply, FACTS based power flow control, etc. [22], [23]. But in the IDM, these inverters are the entities responsible for establishing, regulating and operating the network [22], [30], [34], [38]–[41]. They need to be upgraded from being simple interfacing elements to advanced, grid interactive units with autonomous control.

Migration to IDM introduces additional tasks in the inverter control in the areas of Active and reactive power control and in Grid Synchronization.

- **Active and reactive power control**

The generic goal of any inverter control strategy is to modulate the inverter switching pattern so as to achieve independent control of active and reactive power [5], [22], [23]. They can be broadly classified into (i). Linear Control Strategies with Explicit Modulator and (ii). Nonlinear Control Strategies with Implicit Modulators. Figure 1.4 depicts a flow diagram of the controller classifications. An extensive review of these strategies has been carried out to identify a suitable candidate that can meet the challenges and demands of IDM.

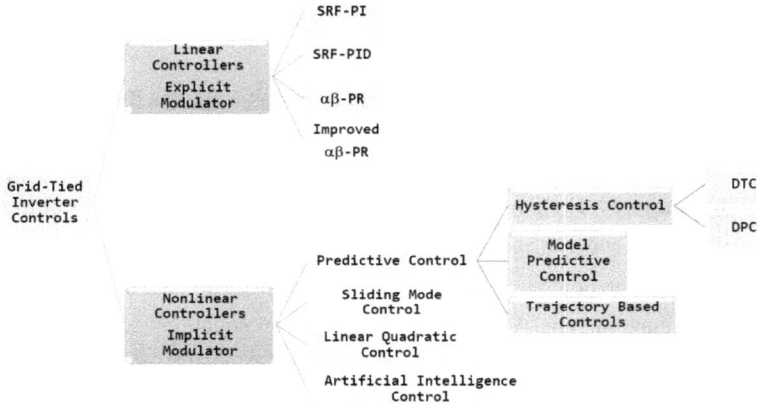

Figure 1.4: Flow diagram of various control strategies for GT inverters

Linear controls approximate the inverter as linear system and employ explicit modulators for switching. They are voltage-oriented schemes with dual control loops with the outer loop performing the voltage control and the inner loop achieves current control. Synchronous Reference Frame control is the most prevalent type of linear control with time tested power tracking accuracy [42]–[49]. They use linear controllers like Proportional Integral (PI) or Proportional-Integral-Derivative (PID) and its variants to reduce steady state errors via feedback loops. Multi-stage reference frame conversions and their computational complexities results poor dynamic response. Further, when working with unbalanced or nonlinear grid signals makes these schemes fail in meeting the reference tracking thus it will be undesirable for IDMs.

Stationary reference frame control (αβ-PR) and its Improved version both work with AC control loops with Proportional Resonant (PR) controller [44], [46], [50]. Because of the elimination of the intense computations for co-ordinate transforms, these schemes offered better dynamic response and improved noise rejection than PI controllers.

Nonlinear controllers on the other hand are robust systems with instantaneous waveform correction, inherent peak current protection, overload rejection, better load dynamics, reduced delay etc. These have implicit modulator, with reduced complexity. Some popular nonlinear control schemes are Predictive Control, Dead Beat Control, Linear Quadratic Control, Artificial Intelligence Control etc. [42]–[49]. Predictive Control use mathematical models of inverters and loads to predict the behavior of the system for a specific time duration. Further, an optimal switching pattern that results in minimum error among the predicted behavior is chosen with the help of specific control criteria [42]–[49]. Hysteresis Scheme with Direct Power Control (DPC) and Direct Torque Control DTC), Model Predictive Control (MPC), trajectory based control are the various subcategories of predictive control. Among them MPC uses the system model for predicting the future behavior [44], [46], [50]. These predictions are further optimized through a specific cost function for attainment of control targets. MPC has multi-objective control capabilities. This means MPC is able to achieve other control targets like system nonlinearities and control constraints simultaneously with the reference tracking [42]–[49]. This multi-objective control feature makes MPC a very attractive choice for implementing advanced functionalities for inverters to work in IDMs.

In addition to active and reactive power control, the grid-tied inverters in IDM are demanded to meet some advanced functionalities viz. multi-mode operation, self- adaptability which are briefed below.

- **Multi-mode operation**

Inverters in IDM are expected to work in multiple operating modes as depicted in Figure 1.5, viz. (i). Grid-Tied, (ii). Grid-Forming and (iii). Grid-Support modes.

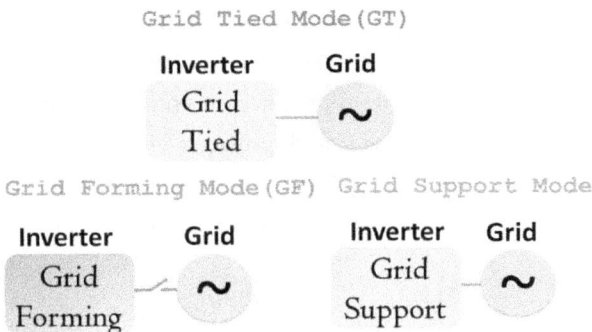

Figure 1.5: Inverter multi-mode operations

➢ Grid-tied mode: In this mode, as depicted in Figure 1.5, the inverter injects/draws predetermined amount of active and reactive power from any RE source to the main grid via the synchronous ties at PCC. The grid synchronization is an essential process in Grid-Tied mode for establishing right magnitude of inverter current representing the referred power. However, synchronization poses new challenges due to uncertain harmonic profiles and new age power quality issues of IDMs. At the same time, IDMs desire inverters with high power tracking accuracy, switching loss reduction to improve efficiency, improved harmonic profile, etc. This demands a multi-objective control which is identified as a research gap.

In this mode often inverters have to work in rectifier mode (bidirectional) so as to facilitate power reversal applicable in energy storage and network power sharing.

➢ Grid-forming mode: Often IDMs work in autonomous mode due to intentional islanding for tariff benefits, improved power quality, generation commitment, etc. or unintentional islanding for ensuring reliability to local customers. Hence the inverters in IDM should be capable of establishing an autonomous local grid with nominal voltage and frequency. Such a control feature is termed as grid-forming control. In grid-forming, multiple inverters must work in parallel adopting power sharing through droop-regulation schemes. This grid-forming capability is an add-on control functionality for a grid-tied inverter when it has to work in the future power grids, because the IDM will claim the liberty to choose to work either in autonomous mode or in grid-tied mode. Additionally, inter-mode transition between the two is yet another mandate. Appending a grid-forming control and mode transition in a grid-tied inverter has emerged as another research gap.

- **Self-adaptability**

Multiple operating modes may demand mode specific performance criteria to be satisfied by the inverters in IDM. For instance, the grid-tied mode demands a strict adherence to stipulated grid codes and standards, while improved efficiency with superior power quality even beyond the mandated grid code may be desired in grid-forming mode. Therefore, the inverter control algorithm should adapt to the mode specific performance indices during multi-mode operation. Absence of such adaptability in the state-of-the-art inverter technology shows the next research gap.

## 1.4.2 Grid synchronization tasks

Research studies on future power quality trends by CIGRÉ and IEEE Task Force, have identified some new age power quality issues like interharmonics, subharmonics (DC and low-frequency components), supraharmonics (components above 1 kHz), etc. which are anticipated to arise in IDMs [44], [46], [50]. The grid voltage signal will no longer remain linear and stationary; instead, it will evolve as non-linear, non-stationary and multi-component signal. If the grid synchronizer is to extract a unit vector template from such non-stationary signals, it requires suitable modifications in the state-of-the-art grid synchronization techniques. Phase Locked Loop (PLL), the common grid synchronizer, is enhanced in the recent past with diverse techniques in order to address a few of the power quality issues like lower order odd harmonics, voltage sag, voltage swell, imbalance, etc. which prevail in today's grid [42]–[49]. But, these advanced PLLs, like Synchronous Reference Frame PLL (SRFPLL), Double Synchronous Reference Frame PLL (dRSFPLL), Dual second order. generalized integrator (DSOGI PLL), Enhanced PLL (EPLL), Moving Average Filter based PLL (MAF PLL), etc. [42], [43], [45]–[47], [49], [51], [52] which are designed to address one specific power quality issue at a time, will underperform when working with multiple power quality issues of IDMs. This research gap has motivated this researcher to look for a multi-component signal decomposition based synchronizer.

## 1.5. Research objectives and methodology

The objectives of the research work are framed as *"Design, develop and implement a MPC scheme in grid-tied inverters for synchronization and power control in IDM with the following features:*

1. *Synchronization using signal decomposition technique to replace the slow responding PLLs for accurate reference generation in non-linear IDM.*

2. *Bidirectional power flow with seamless interoperability between rectifier and inverter modes of operation.*

3. *Grid-forming capability in MPC based grid-tied inverters, and,*

4. *Dynamic Adaptability imparted in MPC for self-adaptive multi-mode operation and mode transition between grid-tied and grid-forming modes. "*

The research methodology involved,

(i) Design of a signal decomposition algorithm to perform as a synchronizer with its parameters tuned to adapt to IDM conditions.

(ii) Sensitivity analysis to study the influence of the MPC design parameters on inverter performance indices.

13

(iii) Development of MPC framework for bidirectional power flow, grid-forming capability, multi-mode operation and smooth mode transition

(iv) Infusion of self-adaptability feature into MPC framework for multi-mode operation and smooth mode transition

(v) Testing of the synchronizer and controller in computer simulation and analysis of its performance under various scenarios, and,

(vi) Validation of the performances of the signal decomposed synchronizer and MPC scheme using hardware in loop.

## 1.6. Thesis outline and organization

This thesis is a compiled report of multiple tasks accomplished to fulfil the research objectives defined in the previous section and is organized into five chapters as follows:

*Chapter.1* encompasses the background and challenges of the next generation IDM with high penetration of RE sources. The research scope and the research gaps identified in this section motivated formulation of four research objectives.

*Chapter.2* describes the conceptualization, design and development of the first research objective - a signal decomposed synchronizer. A signal decomposition algorithm is reoriented to perform as a grid synchronizer for IDM. The proposed synchronizer is tested via simulation and then validated in hardware. Elucidations and key findings are presented in detail.

*Chapter.3* summarizes the work carried out to attain the second research objective - development of MPC based grid-tied inverter for IDM. The key findings of the investigation are presented which form the basis for further work. The section further presents the design of a bidirectional, direct power controlled MPC converter with enhanced dynamics and active rectification. A set of inverter performance indices have been defined for the performance assessment of the proposed control. The results and the analysis are documented in detail.

*Chapter. 4* reports development of a model predictive controlled grid-tied inverter with grid-forming capability. The developed grid-forming capability is validated in simulation and then in hardware. Further, "Dynamic adaptability", a new control feature proposed in this thesis, is discussed at large. The complete design flow, development and performance assessment via MATLAB/Simulink are presented. The testing and validation via Hardware-in-Loop is also included.

*Chapter 5* consolidates the major research findings and suggests some future research directions in this concluding section.

## 1.7. Summary

A broad discussion regarding the emergence of IDM facilitating high RE penetration has been presented. The operational challenges of inverters in IDM from the power control and grid synchronization perspectives are discussed. A review of the literature relevant to the research problem is included and the research gaps in both inverter control and grid synchronization are identified. Lastly, four research objectives and the research methodology are also presented.

# CHAPTER 2

# Design and Development of Variational Mode Decomposition Grid Synchronizer for Inverter Dominated Microgrids

## 2.1 Introduction

The design, development and validation of the proposed signal decomposition based synchronizer for power converters in IDM is presented in this chapter. At first, the capabilities of adaptive time-frequency decomposition techniques to serve as synchronizers are discussed, followed by the elaborations of the identified decomposition algorithms. Further, the simulation and hardware systems used for this study has been introduced and the test results and analysis are presented. Finally, the chapter concludes with a comparison study of the proposed new synchronizer with the state-of-the-art synchronizers along with key research findings.

## 2.2 Signal decomposition approach for grid synchronizer

As discussed in chapter 1, IDMs are foreseen to be the networks with nonlinear characteristics with capricious voltage and frequency profiles. Such non-sinusoidal grid voltage signals are expected to have harmonics and other power quality disturbances overriding on the fundamental frequency component. So, these grid voltage signals can be recognized as non-stationary, Multi-Component Signals (MCS) [14], [53], [54]. Obtaining individual frequency components from a harmonic grid voltage profile is similar to decomposing a composite signal by methods of signal decomposition. Signal decomposition methods are well known for accurate feature extraction of non-stationary signals. This capability, can be utilized with appropriate design reorientation, to help in developing a new generation grid synchronizer. Such a grid synchronizer can exhibit better dynamic response, better harmonic rejection and immunity to transient events because of the inherent characteristics of signal decomposition techniques. Besides extracting the fundamental frequency, a signal decomposition techniques are also capable of simultaneously extracting harmonic frequencies present in the grid signal with acceptable accuracy. Since many applications like harmonic voltage/current measurement [55]–[57], grid impedance estimation [56], [58], [59], transformer saturation detection [60]–[63], neutral current limiting and active power filter control are to handle voltage/current signals contaminated with harmonics [64], [65], there is scope for an effort to reorient a signal processing tool which can perform beyond just grid synchronization.

## 2.2.1 Review on MCS decomposition

MCS decomposition techniques split any non-stationary signal into its constituent frequency components even with complex spectral contents. They are classified as frequency domain, time domain, time-frequency domain and adaptive techniques [14], [53], [54].

Since extraction of fundamental component in the form of time series data is the primary requirement to perform as a synchronizer, signal processing tools like Fourier Transform (FT), Fast Fourier Transform (FFT) and Discrete Fourier Transform (DFT) are no candidates for this application [14]. On the other hand, an improved time domain algorithm like Empirical Mode Decomposition (EMD) has got an edge in handling both stationary and non-stationary signals with equal accuracy, but fails to retrieve the frequency component as time series data [14]. In the recent past adaptive time-frequency decomposition techniques like Wavelet Transform (WT), Empirical Wavelet Transform (EWT), Wavelet Packet Transform (WPT), Variational Mode Decomposition (VMD), etc. emerged [61], [63], [66]–[72]. These techniques demonstrate accurate feature extraction, lesser mode aliasing and avoidance of occurrence of false mode which can effectively meet the requirements of grid synchronizers. These adaptive algorithms have Intrinsic Mode Functions (IMF) which iteratively decomposes the MCS into various modes representing the constituent components as depicted in Figure 2.1. IMFs are special mathematical functions that compute the instantaneous frequency of MCS through Hilbert Huang Transform (HHT) [63], [68], [73].

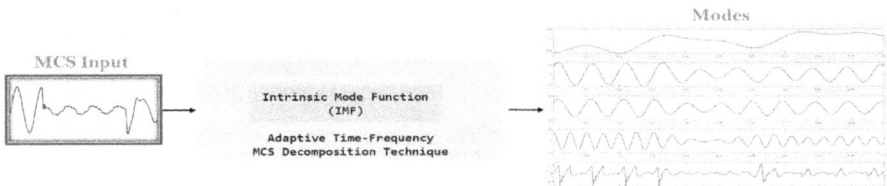

Figure 2.1: Generalized scheme of adaptive time-frequency MCS decomposition algorithms

These new age time-frequency decomposition algorithms offer prospects to meet dual requirements of simultaneous extraction of different frequency components in respective time series. Wavelet Transforms emerged as an improved solution capable to meet the dual

17

requirement of simultaneous extraction of different frequency components as time series, but these are non-adaptive to the signal components [14]. Since the wavelet is a library of functions, the choice of the wavelet must be such prudent that the scaled versions of the wavelet are largely similar to the analyzed signal, in order to ensure good results [58], [59], [74]. Adaptive algorithms like Short Time Fourier Transform (STFT) and Stockwell's Transform (ST), though observed to have accurate feature extraction capability even under small band dynamics, are observed to be inadequate to tackle issues like unknown spectral bands of signals, narrow band frequency deviations, rejection of random transients, etc. [75], [76].

Thus, various adaptive time-frequency decomposition techniques have been reviewed to assess the feasibility as a signal decomposed synchronizer and the results are presented in this chapter.

## 2.2.2 Development of a signal decomposed synchronizer

The signal decomposed grid synchronizer has to be designed to obtain the fundamental frequency component from the grid signal affected by various power quality issues. The desired functionalities of a competent synchronizer are discussed below.

### (i). Extraction of fundamental frequency under unknown spectral composition

Research studies on future power quality issues have identified genres of various grid disturbances like interharmonics, subharmonics and supraharmonics as the key issue on non-linear grids [14]. The signal decomposed grid synchronizer has to be designed to tackle all these new age power quality issues anticipated in the IDM and some of these challenges are briefed below:

- *Even harmonics:* CIRÉD recognized Power electronics fed wind energy systems as a significant source of even harmonic emissions [14]. The interfacing converters often operate at switching frequencies of 2 kHz or higher for efficient power control. This results in even harmonics of the orders of 36, 38, 40 etc. to be injected on to the grid. Such even harmonics are uncommon in legacy grids.
- *Interharmonics:* Interharmonics is defined by IEEE as "*a non-integer frequency component of a periodic quantity that is not an integer multiple of the fundamental frequency*" [77]. New age loads like LED lamps, induction devices and arc furnaces also draw non-linear harmonic currents due to interharmonics [14]. Such non-characteristic

harmonic frequencies may cause other problems like resonance with capacitive compensators, faults in motor drives, etc.

- **Subharmonics:** The frequency components whose frequencies are less than the fundamental grid frequency are called as subharmonics or quasi-DC. Low frequency interharmonic emissions are a major source of subharmonics along with misfired power electronics [14]. Subharmonics can cause saturation of transformer core, reduction of the life of AC motors, resonance in turbine shafts, etc.

- **Supraharmonics:** Grid voltage harmonics at inverter switching frequencies or its multiples (2 kHz- 15 kHz) are named as supraharmonics [14]. Self-commutated power converters, electronic ballast, new age lighting loads etc. inject supraharmonics. They can cause resonance and interference with capacitive energy storage systems or compensators.

**(ii). Fundamental frequency deviations**: Due to low inertia, the fundamental grid frequency deviates in a band of about ± 2Hz from its nominal value [14]. The synchronizer which adopts signal decomposing algorithm must possess high levels of feature extraction and feature classification accuracies so that they can track any deviation in the fundamental frequency.

**(iii). Phase jump**: Increased use of underground cables or overhead lines are the causes of a significant deviation in the phase angle of grid voltage, called as phase jumps [14]. The new age grid synchronizer is expected to follow any such phase deviation in the fundamental frequency grid voltage component accurately.

**(iv). Random transients**: Various forms of transient power quality disturbances like multiple zero crossings, short duration magnitude surges, etc. are other power quality disturbances clustered under low inertial systems [14]. These transients cannot be predicted and are a completely random phenomenon. The grid synchronizer is expected to reject such transients and extract the fundamental frequency component with utmost accuracy.

**(v). Online feature extraction time**: The time required for fundamental frequency extraction will be an important factor which can give good dynamic response of the synchronizer. This includes the minimum observation window size and the execution time for the algorithm. The signal decomposition technique intended to be used in synchronizer must extract the fundamental frequency component from a short observation window within an execution time less than the time period of the fundamental grid frequency.

19

## 2.3 Feasibility study of decomposition algorithms for synchronizer application

A detailed review of the various adaptive signal decomposition techniques has been done. Based on the information available in the literature correlated to the requirements of the grid synchronizer, a feasibility analysis has been carried out to assess the suitability of each of these techniques to be designed as synchronizer.

### 2.3.1 Elucidations and findings of the feasibility study

(i). **EMD** is an adaptive time-frequency technique in which IMF is formulated to decompose MCS based on local maxima and local minima amplitude [55]–[57], [61], [62], [78]. The decomposed sub-signals are then grouped into modes with the same values of amplitude. EMD preserves time information but with a trade off on loss of selective frequency extraction [61], [62]. It also fails to retrieve the frequency component as a time series for the components of MCS, which makes this algorithm inferior to perform as a grid synchronizer.

(ii). **STFT** and **ST** were also effective in segregating the linear signals, but have poor accuracy with non-linear signals [55], [57], [65], [75], [76], [78]–[81].

(iii). **EWT** combines the filtering technique of WT and amplitude segregation features of EMD [55]–[57], [67], [70], [71], [78], [82]–[84]. Hence, EWT is proven to exhibit better decomposition accuracies for non-stationary and highly noisy MCS compared to WT and EMD. Moreover, its unique ability to segregate the constituent frequency components even with the presence of large number of harmonic frequencies has made it very popular in fault identification and power quality event characterization [55], [82], [85]. EWT was also observed to have accurate feature extraction capability even under small band dynamics, narrow band frequency deviations, random transients, etc. and thus has potential to work as grid synchronizer.

(iv). **VMD** is reportedly competent in addressing the issues like (i) simultaneous localization of time and frequency, (ii) lack of mathematical foundation; (iii) decomposition for transient events, [68], [69]. VMD divides MCS into time series components centered around a single frequency or a mode with extremely small sidebands. So, the constituent frequencies in a composite signal can be extracted as multiple time series subcomponents. However, fitting various VMD optimization parameters while working with signals possessing diverse composition of spectral bands is the key to ensure crisp decomposition with commendable accuracy [66], [68], [69], [88], [89]. For grid synchronizer, the optimization parameters in VMD should be tuned to extract the fundamental frequency from a non-stationary IDM grid voltage signal in its lower modes. The high frequencies if present in the grid signal will be pushed to higher modes can be made to ignore by the synchronizer. VMD also accommodates spectral side bands which can be utilized to capture the expected deviations in the fundamental frequency of the electric grid. These features make VMD an ideal candidate to serve as the grid synchronizer.

From this feasibility study VMD and EWT have been identified as apt decomposition techniques which can be utilized for synchronization application. Therefore, both VMD and EWT synchronizers have been designed, developed, tested and presented in this thesis.

## 2.4 Design and development of VMD grid synchronizer

### 2.4.1 VMD algorithm

VMD algorithm performs singular value decomposition of a MCS described by $f$ into finite number of discrete sub-signals or modes, $u_k$, where $k= 1, 2.....K$, subject to constraints [66], [68], [69], [88], [89]. Once decomposed, the Fourier transforms, $f_k$ of $u_k$ are restricted around a central frequency, $\omega_k$, with a narrow bandwidth, $\Delta\omega_k$. Every $u_k$ will be retrieved as a time series signal, with no overlap between consecutive $\omega_k$. This is achieved by minimizing the sum of bandwidths in each $u_k$ such that upon recombination of these modes, the original signal will be reproduced in its actual form without loss of any information. This is the constrained variational problem solved in VMD and is expressed as,

$$min_{u_k\omega_k} \sum_k \Delta\omega_k \quad such\ that \quad \sum_k u_k = f \qquad .........(2.1)$$

Figure 2.2 depicts the process flow block diagram of the VMD synchronizer. The first step of this process is to obtain the frequency spectrum of '$f$'so as to decompose it into modes at every constituent central frequency. For this purpose, an analytic function $u_k^A (t)$ based on the mode $u_k (t)$ is constructed by finding the Hilbert Transform [73] of the component. The Hilbert Transform computes the instantaneous frequency of the mode and is obtained as the complex conjugate of the mode, represented as,

$$u_k^A (t) = u_k(t) \left( \delta(t) + \frac{j}{\pi t} \right) \qquad .........(2.2)$$

Then, every $u_k^A$ (t) for k= 1,2............. K, is demodulated to have a base band by multiplying it with a complex exponential $e^{j\omega t}$ to shift its constituent frequency as its central frequency, $\omega_k$.

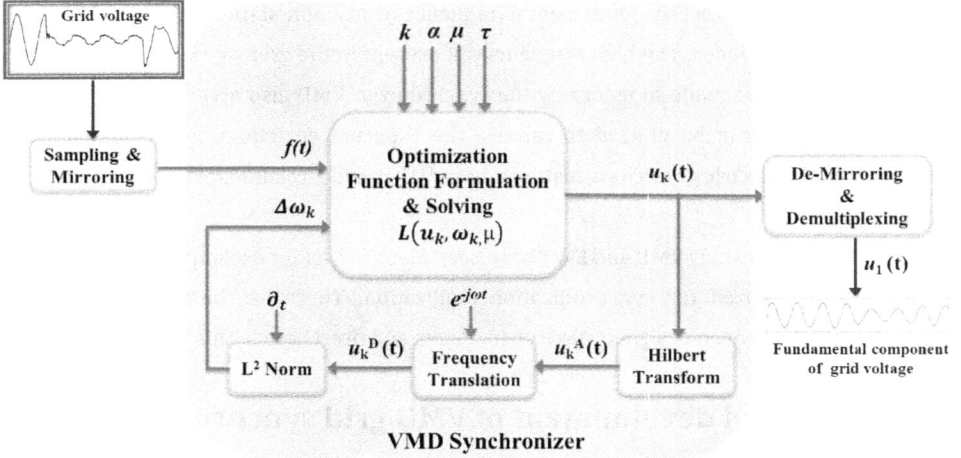

Figure 2.2: Process flow block diagram of the VMD Synchronizer

This gives $u_k{}^D$, the frequency-translated form of $u_k{}^A(t)$, expressed as,

$$u_k^D(t) = u_k^A(t)e^{-j\omega_k t} \quad .........(2.3)$$

Now, the bandwidth, $\Delta\omega_k$, for every mode can be obtained as the squared $L^2$ norm of the gradient, calculated as,

$$\Delta\omega_k = \left\| \partial_t \left(u_k^D\right) \right\|_2^2 \quad .........(2.4)$$

Thus, the constrained variational problem given in equation (2.1) can be re-written as,

$$min_{u_k\omega_k} \sum_k \left\| \partial_t \left[ \left(\delta(t) + \frac{j}{\pi t}\right) * u_k(t)e^{-j\omega_k t} \right] \right\|_2^2 \; Such \; that \; \sum_k u_k = f \quad .........(2.5)$$

This constrained optimization problem shown in equation (2.5) is next converted into unconstrained optimization problem by using augmented Lagrangian method as,

$$L\left(u_k, \omega_k, \mu\right) = \alpha \, min_{u_k\omega_k} \Sigma_k \parallel \partial_t(u_k^D) \parallel_2^2 + \parallel f - \Sigma_k u_k \parallel_2^2 + \parallel \mu \, (f - \Sigma_k u_k) \parallel \quad .........(2.6)$$

where, $\alpha$ is the balancing parameter of the data-fidelity constraint and $\mu$ is the Lagrangian multiplier. The solution to the optimization problem in equation (2.6) is now found as the saddle point of the augmented Lagrangian in sequence of iterative sub-optimizations called alternative direction method of multipliers (ADMM), given by

$$u_k^{n+1} = \frac{f - \Sigma_{i<k} u_i^{n+1} - \Sigma_{i>k} u_i^n + {\mu^n}/{2}}{1 + 2\alpha\left(\omega - \omega_k^n\right)^2} \quad .........(2.7)$$

$$\omega_k^{n+1} = \frac{\int_0^\infty \omega |u_k^{n+1}(\omega)|^2 d\omega}{\int_0^\infty |u_k^{n+1}(\omega)|^2 d\omega}$$

.........(2.8)

$$\mu^{n+1}(\omega) = \mu^n(\omega) + \tau(u(\omega) - \sum_k u_k^{n+1}(\omega)) \quad .........(2.9)$$

where, n is the iteration step, τ is the tolerance of convergence, $u_k^{n+1}$ and $\omega_k^{n+1}$ are the updated $u_k$ and $\omega_k$ components for each iterative step. Equation (2.7), by which each mode is extracted, resembles as the *Wiener filter function* typically used for de-noising any MCS [66]. The components obtained by VMD in turn depend upon parameters $k, \alpha$ and $\mu$ of equation (2.6) and 'τ' assigned to terminate the optimization.

In the proposed grid synchronizer application, VMD is intended to capture only the fundamental frequency component rather than signal reconstruction; so $\mu$ in equation (2.6) is made zero [68], [69], [88], [89].

Therefore, equation (2.6) becomes,

$$L(u_k, \omega_k) = \alpha \min_{u_k, \omega_k} \sum_k \left\| \partial_t \left[ u_k^p \right] \right\|_2^2 + \left\| f - \sum_k u_k \right\|_2^2$$

.........(2.10)

Now, equation (2.10) is optimized to suit the requirements of grid synchronization; the test result of the same is presented in the following section.

## 2.4.2 Design consideration of VMD parameters

The VMD algorithm demands the number of modes, $k$, as equal to the number of frequencies present in any spectral band of MCS [68], [88]. When this number is not known in an MCS, then assigning an arbitrary value of $k$ reduces the accuracy of extraction. As the future power grids are expected to have grave harmonic footprints of voltages and currents with random spectral contents of different amplitudes, maintaining high accuracy of extraction needs appropriate choice of $\alpha$ and $k$ [68], [90].

The prime objective of grid synchronizer is a fine separation of the fundamental frequency; so, the Weiner filter function [68], [69] is made narrow and made to center around the grid frequency with a small sideband so that deviations in it can also be captured without fail. Higher values of $\alpha$ yield a narrow filter bandwidth which eventually does a finer frequency separation, but it will fail to capture minor deviations of the principal or the center frequency. In contrast, lower values of $\alpha$ widen the filter bandwidth, thereby facilitating the capture of the deviated frequencies, but provide a worse separation [68]. However, extremely narrow bandwidths may not recognize the minor fundamental frequency shifts inherently present in electric grids. So, a mid-range value of $\alpha$ is chosen for the proposed application.

Another vital concern for any sampled signal processing system is the input window size; higher window size allows a reliable decomposition process, but yields poor dynamic response, which is intolerable in synchronizers. Generally, the window size is chosen as the full cycle period of the lowest frequency component to be fetched [88]; but it makes the synchronization sluggish while tracking low frequencies.

VMD has proved to reliably decompose any MCS with a minimum window size equal to quarter cycle period of the frequency to be extracted, which is at par with that of improved PLL structures [91]. Therefore, a minimum window size equal to quarter cycle period of fundamental frequency is chosen in this work.

# 2.5 Performance evaluation of proposed VMD synchronizer

The performance evaluation of the proposed VMD synchronizer is carried out through assorted test cases, formulated specially to include all the possible power quality events foreseen in IDMs. Also, multiple power quality events have been created in some test cases to represent the worst-case scenarios. In each case, the synchronizer is designed to extract the fundamental frequency signal at one of its low frequency modes with highest possible accuracy.

## 2.5.1 Testing in simulation platform

The implementation of the VMD synchronizer and visualization of its performance have been accomplished with the simulation platform of MATLAB/Simulink. The test signals have been obtained through separate simulations with appropriate Simulink libraries and stored as time series data which could further be applied to the VMD algorithm. The algorithm has been coded as a MATLAB script and the results have been obtained in the *scope* tool of Simulink. The extracted fundamental frequency signals have been then analyzed and assessed for the feature extraction accuracy.

**(a). Formulation of test cases**

Four categories of test scenarios have been considered to formulate 13 test cases.

- The first category represented the present-day grid signals with lower order harmonics, occasional voltage sag/swell, distortions, etc. These test signals have well known spectral composition which are standard and are readily generated with

MATLAB/Simulink components.

- The second category provided signals anticipated in future power grids based on the *CIGRÉ Study Committee C4* report on "*Power quality and EMC issues with future electricity network*" [92]. The major power quality concerns listed in the *CIGRÉ* report, which includes interharmonics, DC components, subharmonics (quasi-DC) and supraharmonics, are carefully tailored into the test cases considered for the present study.

- Real time field measurements collected via digital storage oscilloscope (Tektronix TBS1062) have also been used for the study as the third category of test signals to validate the performance of the proposed synchronizer with real world data.

- Data sets taken from IEEE 1159.2 taskforce for power quality event characterization formed the fourth category of test signals [92]. This IEEE data set can characterize multiple power quality events which are expected to be very common in the future IDM.

## (b). Testing of proposed grid synchronizer

An input voltage signal, $v_{(ph)x}$, *(where x indicates the test case serial number)* represents the grid voltage and the synchronizer is expected to extract its fundamental frequency at its first mode, $u_1$.

- **Testing with the present-day grid signals**

Case 1a: Lower order harmonics

This case represents the present-day grid voltage with four odd harmonics ranging from 5th to 13th order. The input signal formulated for this case is shown in Figure 2.3.(a) which can be described as,

$$v_{(ph)1a} = \sqrt{2} \times 240 \, sin \, sin \, (2\pi 50t) + \sqrt{2} \times 48 \, sin \, (2\pi 250t) + \sqrt{2} \times 24 \, sin \, (2\pi 350t)$$
$$+\sqrt{2} \times 12 \, sin \, (2\pi 550t) + \sqrt{2} \times 12 \, sin \, (2\pi 750t)$$

The data set $V_{(ph)1a}$ in this case can typically be measured at the front end of a modern LED lamp [14].

Case 1b: A voltage signal with Line notching is considered in $v_{(ph)1b}$ and the same is shown in Figure 2.4.(a) which is the input to the VMD synchronizer in this case.

- Testing with Signals anticipated in future IDM

Case 2: Supraharmonics

This test case brings in a futuristic grid condition, wherein supraharmonics in the range from 2 to 150 kHz will be significant. The input, $v_{(ph)2}$, presented in Figure 2.4.(b) is emulated to represent a grid voltage overridden by 5th order lower harmonics and 39th order supraharmonics. The signal can be represented as,

$$v_{(ph)2} = \sqrt{2} \times 240\ sin\ (2\pi 50t) + \sqrt{2} \times 48\ sin\ (2\pi 250t) + \sqrt{2} \times 24\ sin\ (2\pi 1950t)$$

Cases 3 & 4: Voltage sag and swell

The Case 3 and Case 4 input signals, $v_{(ph)3}$, and $v_{(ph)4}$, emulate transient events of voltage sag and voltage swell respectively. A voltage swell and sag with 10% variation from the nominal value has been introduced at the instant of 0.5s for a period of 0.05 s, with an overriding 5th harmonic. The test input for voltage sag has been presented in Figure 2.5.(a). The task here is to test the transient response in magnitude.

Cases 5 & 6: Phase jump

It introduces a deviation in the phase angle of the fundamental frequency component. The phase angle deviations introduced here are +10 ° for $v_{(ph)5}$ in Case 5 and -10 ° for $v_{(ph)6}$ in Case 6, each at the instant of 0.5s for a period of 0.05 s; the task is to extract the fundamental frequency and assess the phase tracking accuracy. The test inputs for cases 5 and 6 have been presented in Figure 2.5.(c) and (d).

Case 7: Frequency change

The input signal $v_{(ph)7}$ of this test case represents a grid frequency deviation [5], [18]. In this test case, a frequency deviation of ±0.5 Hz from its nominal value is introduced at the instant of 0.5s for a period of 0.05s, with the objective of capturing the frequency information through VMD synchronizer using multiple sampling frequencies of 10 kHz, 20 kHz, and 50 kHz. VMD is an iterative, yet adaptive, optimization technique wherein sampling rates decides the accuracy of input signal capture. Thus, analyzing the correlation between sampling rates and frequency tracking capability in this case draws significance.

Case 8: Unbalanced voltage

A magnitude imbalance of ±10% is introduced as a voltage rise in phase 'a' at the instant of 0.5s for a period of 0.1s, followed by a voltage fall.

Case 9: Interharmonic emissions

This test case is formulated with an interharmonic and a supraharmonic, both overriding the fundamental, as shown in Figure 2.7.(a). An exclusive tracking task is set for this case to retrieve both the fundamental and the interharmonic, for synchronization and islanding detection respectively. The test input signal, $v_{(ph)9}$ comprises the nominal fundamental voltage, a 10% interharmonics at 75 Hz and a 10% supraharmonics of 1950 Hz [18], expressed as,

$$v_{(ph)9} = \sqrt{2} \times 240\ sin\ (2\pi50t) + \sqrt{2} \times 24\ sin\ (2\pi75t) + \sqrt{2} \times 24\ sin\ (2\pi1950t)$$

Case 10: Surge ride through

The VMD synchronizer is tested in this case for surge ride through capability. A voltage transient of 10% above the nominal value is introduced to the input voltage, $v_{(ph)10}$, at the instant of 0.5s for a period of 0.005s.

Case 11: Zero crossing distortions

This case is formulated with the input, $v_{(ph)11}$, that has multiple zero crossing over a period of 0.0005s starting at the instant of 0.1s, as shown in Figure 2.8.(a). The task is to extract the fundamental frequency identifying the right zero crossing while ignoring multiple ones.

- **Testing with real field data**

Case 12: Distribution Feeder Data

This case uses the real field data collected from the university campus distribution feeder which is rich in harmonics owing to non-linear loads like UPS, rectifiers and computer lab. The voltage data, $v_{(ph)12}$, is collected using digital storage oscilloscope (DSO), (Tektronix TBS1062), and Power Quality Analyzer, (Fluke LM 434), simultaneously at various nodes of a LT feeder. The time series voltage data from DSO form the signal input to VMD and the power quality analyzer data is used as a benchmark for validation. The objective of this test case is to check the fidelity of extraction when working with signals of arbitrary spectral contents, so that a generic VMD parameterization can be followed for practical implementations. Figure 2.9 presents the data.

- **Testing with signals from IEEE Task Force 1159.2**

Case 13: IEEE Task Force 1159.2 Data

Case 13 is formulated with five signals bearing multiple power quality issues, including sag, swell, frequency jump, notching, spikes, arbitrary distortions, etc., chosen from the IEEE Task Force 1159.2 repository [92]. The test signal is shown in Figure 2.10.

## 2.5.2. Performance analysis and observations

The performance of the synchronizer is analyzed using few error indices based on the grid connection standards. For example, the signal level extraction accuracy is one such index in grid current control applications in which cycle level values of the signals are computed. High accuracy at cycle level can maintain strict reference tracking of active power and power factor, besides offering reduced circulating currents, reliable synchronization, etc. The RMS error is meant for the signal shape, which represents the deviation between the retrieved fundamental component and the actual fundamental component of the input signal. Less deviation in each sample helps in delivering inverter current of pure sine waveform which can meet, or even exceed, the harmonic standards for grid connection.

Thus, based on the grid connection standards, two statistical error indices, viz. (i) the absolute errors in the magnitude, phase and frequency, '$\varepsilon$', and (ii) the RMS error of the sampled data, are used to assess the accuracy of extraction. The absolute error in magnitude, $\varepsilon_v$, $phase$, $\varepsilon_\theta$, and frequency, $\varepsilon_f$, computed as,

$$\varepsilon_v = \left(\frac{v^*_{rms}-v_{rms}}{v^*_{rms}}\right) \times 100 \quad .........(2.11)$$

$$\varepsilon_\theta = \left(\frac{\theta^*-\theta}{\theta^*}\right) \times 100 \quad .........(2.12)$$

$$\varepsilon_f = \left(\frac{F^*-F}{F^*}\right) \times 100 \quad .........(2.13)$$

where $v^*_{rms}$ and $v_{rms}$ are the rms values and $\theta^*$ and $\theta$ are the phase angle values of the nominal grid voltage and the extracted fundamental component, and, $F^*$ and $F$ are the nominal grid frequency and the frequency of the extracted fundamental component respectively. The root mean square error can be computed as,

$$RMSE = \sqrt{\sigma^{*2} - \sigma^2} \quad .........(2.14)$$

where $\sigma^*$ and $\sigma$ are the standard deviations of the sampled values of the nominal grid voltage and the extracted fundamental frequency component respectively.

The input signals pertaining to each test case and the corresponding extracted time series ($u_k$) are presented in the plots shown in Figure 2.3 to 2.11. The frequency extraction in each case has the objective to retrieve the required '$k^{th}$' harmonic as mentioned in the test case description. The computed error indices for cases 1 to 13 are consolidated and presented in Tables 2.1, 2.2 and 2.3.

The applied input, $v_{(ph)1a}$ and its segregated VMD modes for case.1a. are presented in Figure 2.3.(a), where the extracted fundamental component is shown as $u_1$; the higher harmonics segregated as $u_2$ to $u_5$ are presented in Figures 2.3.(b) to (e). VMD synchronizer has exhibited

28

an accurate frequency segregation with highest resolution at all modes, but it has increasing errors in magnitude and phase at higher frequency modes. This is further ascertained by the frequency spectra in Figure 2.3.(f) and (g).

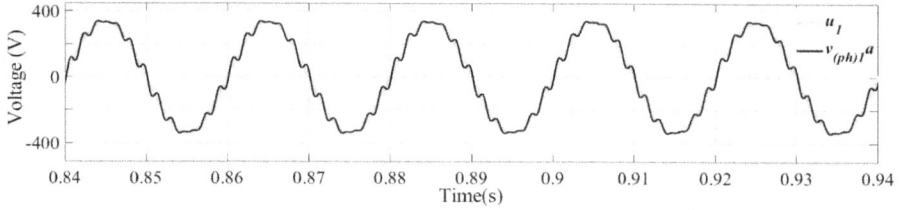

(a)

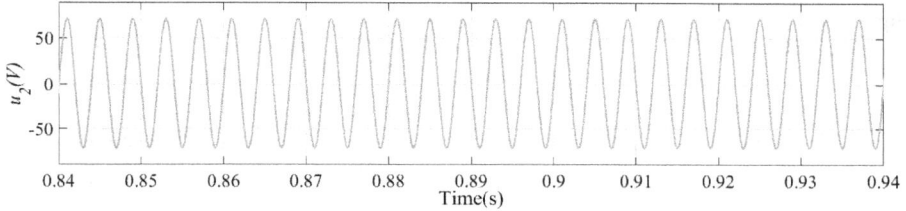

(b)

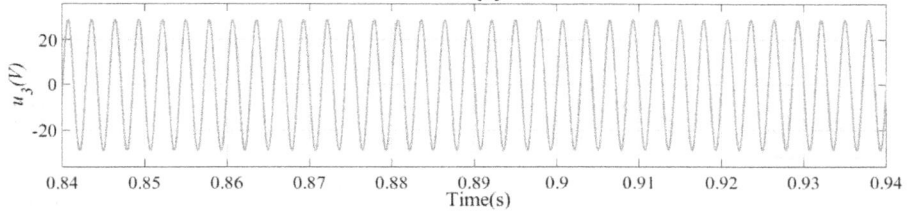

(c)

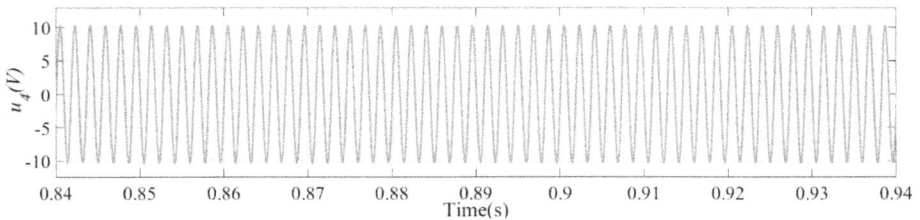

(d)

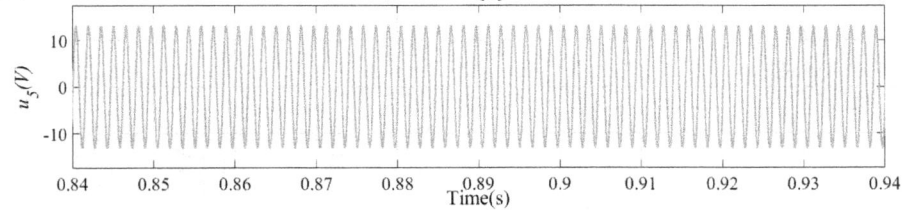

(e)

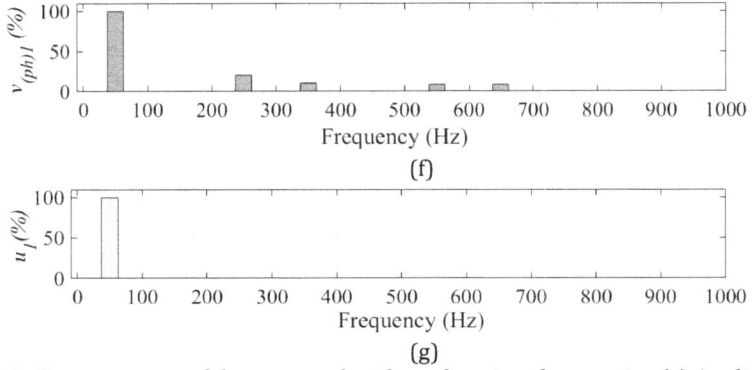

Figure 2.3: Time response of the proposed grid synchronizer for case 1.a, (a) Applied $v_{(ph)1a}$ and Extracted mode $u_1$; (b) to (e) Extracted modes $u_2, u_3, u_4, u_5$ and (f) to (g) Frequency spectrum of applied $v_{(ph)1a}$ and $u_1$.

The spectrum of $u_1$ has only the fundamental frequency component, while the input signal spectrum shows the entire frequency components of the applied signal. This validates the customized design parameter of VMD for accurate feature extraction at the fundamental frequency mode.

TABLE 2.1 EXTRACTION ERROR COMPUTED FOR TEST CASES 1 TO 11

| Case | Absolute error in waveform, (ε (%) | | | | RMS error in sampled data | | | |
|---|---|---|---|---|---|---|---|---|
| | Frequency | Magnitude | | Phase | | Magnitude | | Phase |
| 1.a. Lower order, odd harmonics | 0 (for all frequencies) | $v(u_1)$ | 0 | $\angle(u_1)$ | 0.056 | $v(u_1)$ | 0.6875 | $\angle(u_1)$ | 0.0182 |
| | | $v(u_2)$ | -3.69 | $\angle(u_2)$ | 0.111 | $v(u_2)$ | 1.657 | $\angle(u_2)$ | 0.689 |
| | | $v(u_3)$ | 14.65 | $\angle(u_3)$ | 0.139 | $v(u_3)$ | 3.569 | $\angle(u_3)$ | 0.9674 |
| | | $v(u_4)$ | 15.24 | $\angle(u_4)$ | 2.47 | $v(u_4)$ | 4.632 | $\angle(u_4)$ | 1.563 |
| | | $v(u_5)$ | 19.53 | $\angle(u_5)$ | 3.856 | $v(u_5)$ | 5.236 | $\angle(u_5)$ | 1.968 |
| 1.b. Grid voltage at pcc with line notching 2.supraharmonics | 0 | $v(u_1)$ | 0 | $\angle(u_1)$ | 0.056 | $v(u_1)$ | 0.6875 | $\angle(u_1)$ | 0.0182 |
| 3. Voltage sag 4. Voltage swell 5. Phase jump (lag) 6. Phase jump (lead) 8. Voltage unbalance | 0 | $v(u_1)$ | 0 | $\angle(u_1)$ | 0.056 | $v(u_1)$ | 0.6995 | $\angle(u_1)$ | 0.0182 |
| 7. Frequency jump | 0.01 | $v(u_1)$ | 0 | $\angle(u_1)$ | 0.056 | $v(u_1)$ | 0.6995 | $\angle(u_1)$ | 0.0182 |
| 9. Interharmonics | 0 | $v(u_1)$ | 0 | $\angle(u_1)$ | 0.056 | $v(u_1)$ | 0.6875 | $\angle(u_1)$ | 0.0182 |
| | 0 | $v(u_2)$ | 5.21 | $\angle(u_2)$ | 0.056 | $v(u_2)$ | 0.8965 | $\angle(u_2)$ | 0.0281 |
| 10. Voltage surge | 0 | $v(u_1)$ | 0 | $\angle(u_1)$ | 0.056 | $v(u_1)$ | 0.6875 | $\angle(u_1)$ | 0.0182 |
| 11. Multiple zero crossing | 0 | $v(u_1)$ | 0 | $\angle(u_1)$ | 0.1111 | $v(u_1)$ | 0.7465 | $\angle(u_1)$ | 0.0281 |

30

The steady state time response in processing grid voltage with line distortion is depicted in Figure 2.4.(a), in which the fundamental component has been fetched with clear rejection of low frequency harmonics.

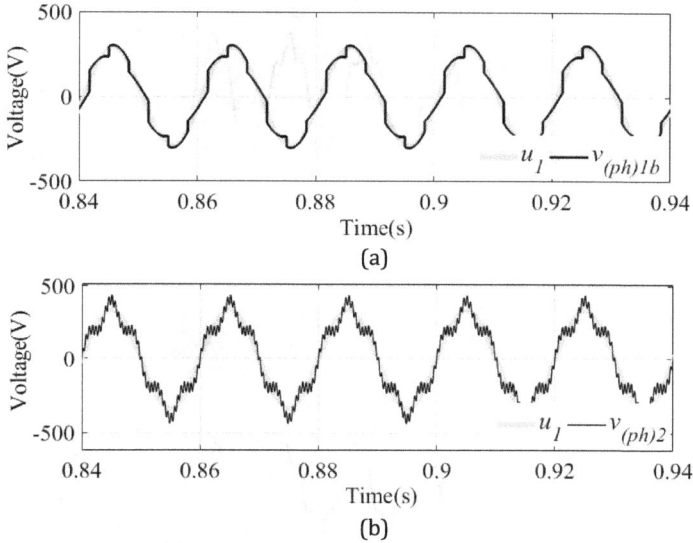

(a)

(b)

Figure 2.4: Time response of the proposed grid synchronizer, (a) Applied $v_{(ph)1b}$ and extracted $u_1$ in Case 1.b and (b). Applied $v_{(ph)2}$ and extracted $u_1$ in Case 2.

The response of the VMD synchronizer for grid voltage signal with supraharmonic component has been depicted in Figure 2.4.(b). The extracted fundamental component is seen to have an error of 0%, 0% and 0.056% in frequency, magnitude and phase respectively. The supraharmonics are observed to have shifted to the higher modes of the VMD during the decomposition. Thus, VMD synchronizer has been seen to exhibit a clear rejection of supraharmonics.

Responses of the VMD synchronizer for cases 3 and 4 (voltage sag and voltage swell) as well as case 5 and 6 (phase jump (lag) and phase jump (lead)) are depicted in Figure 2.5.(a) to (d). The performance indices of the synchronizer response for these cases have been presented in Table 2.1. It shows the that the synchronizer is able to detect both the transient events the sag and the phase jump (lag) with commendable tracking speed. The extracted magnitudes of 325 V and 292.5 V in the consecutive peaks, shown in the insert of Figure 2.5.(a), shows how the synchronizer follows the sag. Similarly, the phase jump introduced at the instant of 0.5 s is found captured by the synchronizer as seen in the insert of Figure 2.5.(c).

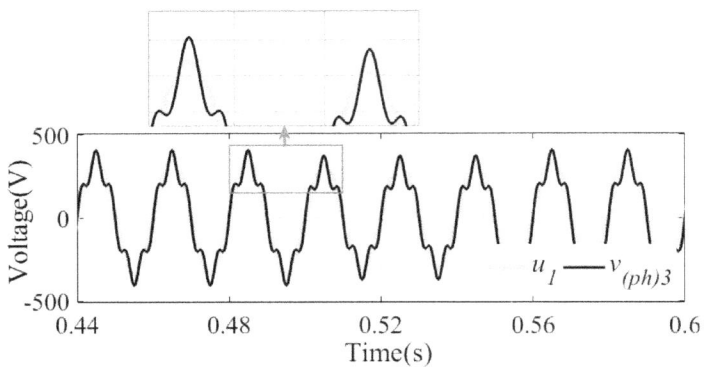

(a)

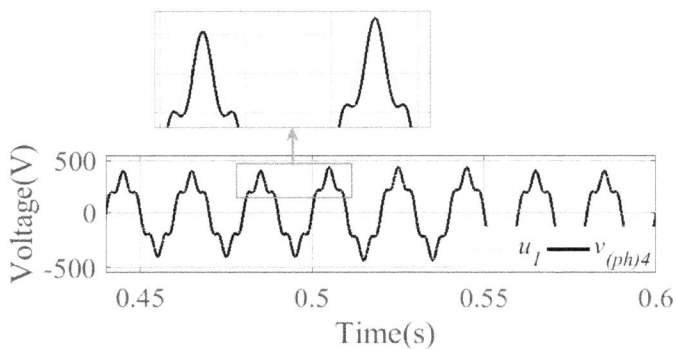

(b)

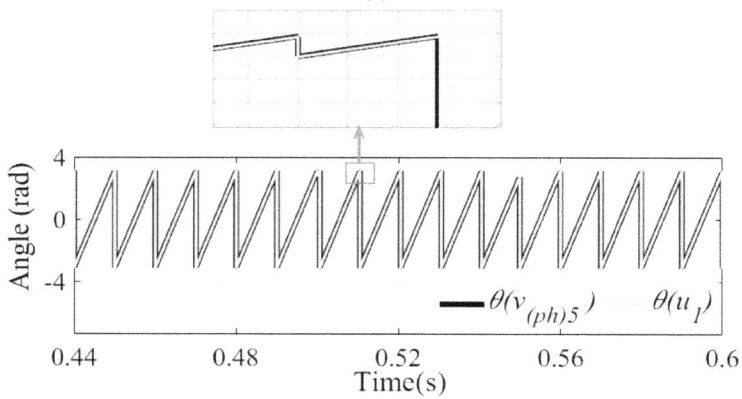

(c)

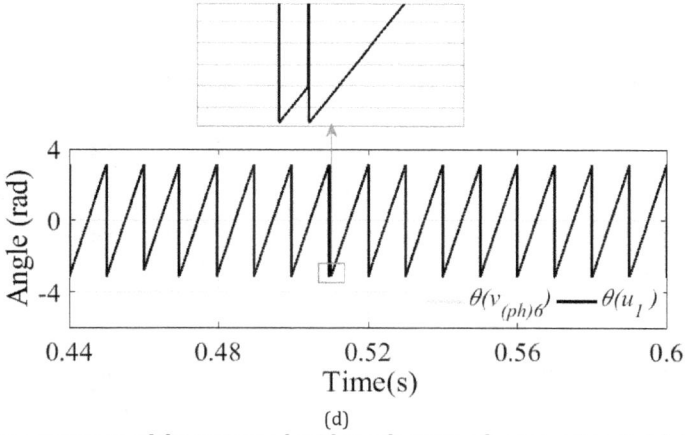

(d)

Figure 2.5: Time response of the proposed grid synchronizer for cases 3 to 6, (a) Applied $v_{(ph)3}$ with a sag at the instant of 0.5s to 0.55s and the extracted $u_1$ in Case 3, (b) Applied $v_{(ph)4}$ with a swell at the instant of 0.5s to 0.55s and the extracted $u_1$ in Case 4, (c) Applied $\theta_{(v(ph)5)}$ with a phase jump at the instant of 0.5s to 0.55s and extracted $\theta_{(u1)}$ in case 5 and (d) Applied $\theta_{(v(ph)6)}$ with a phase jump at the instant of 0.5s to 0.55s and extracted $\theta_{(u1)}$ in case 6.

The results for frequency jump are presented with two additional performance indices, viz. instantaneous frequency ($f_{ist}$) and frequency transition time ($\Delta_{trans}$) [73], [94], to evaluate the competency of the synchronizer in capturing frequency deviations. $f_{ist}$ is computed as time derivative of the instantaneous phase using Hilbert Transform [73], [94], while $\Delta_{trans}$ is the minimum time required for $f_{ist}$ to settle at its final value following a deviation. Figures 2.6. (a) to (c) show the $f_{ist}$ and $\Delta_{trans}$ of the extracted $u_1$ of the case 7 sampled at three different sampling frequencies.

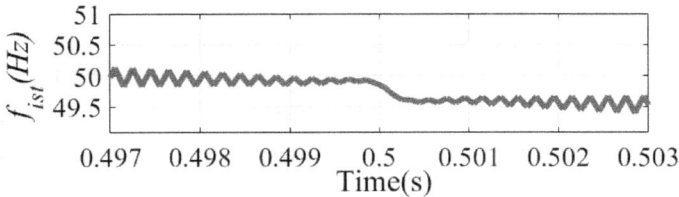

(a)

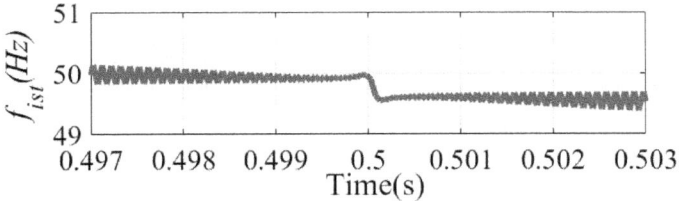

(b)

33

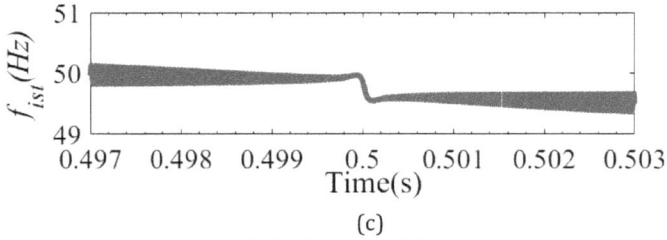

Figure 2.6: Instantaneous frequency ($f_{ist}$) of extracted fundamental component magnified at the instant of transition for Case 7 sampled at (a) 10 kHz, (b) 20 kHz and (c) 50 kHz.

In Figure 2.6.(a) to (c), the frequency prior to the transition at 0.5 s is observed to be 50 Hz while after the transition is observed to be 49.5 Hz. But, $\Delta_{trans}$ shows different values as influenced by the sampling rates. Transition times of 2ms, 1.8ms and 1.4ms are observed for sampling frequencies of 10, 20 and 50 kHz respectively. It infers that higher sampling frequencies result in lower transition times, which in turn provide better dynamic tracking ability for the synchronizer. Table 2.2 presents the frequency settling times of various modified and advanced PLL schemes reported in literature.

TABLE 2.2 COMPARISON OF TRANSITION TIME

| Advanced PLL algorithm/ method | dqPLL | αβPLL | Ddsrf PLL | DSOGI PLL | Dαβ PLL | FPD dαβPLL | Adaptive FPD dαβPLL | MAFPLL |
|---|---|---|---|---|---|---|---|---|
| $\Delta_{trans}$ (ms) | 75 | 101 | 70 | 48 | 200 | 124 | 53 | 200 |
| Advanced PLL algorithm/ method | EPLL | EPMAF PLL | EPMAF PLL Type 2 | αβ EPMAF PLL | MSHDC PLL | DNαβ PLL | EMAF PLL | VMD Synchronizer |
| $\Delta_{trans}$ (ms) | 132 | 60 | 58 | 132 | 40 | 60 | 132 | 2 |

The settling time of the VMD synchronizer tested and presented here is found low when compared to the PLL structures like Synchronous reference frame PLL (dqPLL), stationary reference frame PLL (αβPLL), decoupled double synchronous reference frame (ddSRFPLL), second-order generalized integrator PLL (SOGIPLL), double stationary reference frame PLL (dαβPLL), fractional phase divider double stationary reference frame PLL (FPDdαβPLL), advanced fractional phase divider double stationary reference frame PLL (Advanced FPDdαβPLL), moving average filter PLL (MAFPLL), enhanced PLL (EPLL), enhanced moving average filter PLL (EMAFPLL), stationary reference frame enhanced PLL (EPLL), enhanced moving average filter PLL (EMAFPLL), enhanced phase error estimation moving average filter PLL (EPMAFPLL), stationary reference frame enhanced phase error estimation moving average

34

filter PLL (αβEPMAFPLL), multi sequences harmonic decoupling network based PLL (MSHDCPLL) and decoupling network stationary reference frame PLL (DNαβPLL) [46], [50], [51], [55], [91], [95]–[97].

Results of case 8, corresponding to unbalance is only in Table 2.1. The test results of simultaneous extraction of the fundamental component and the closely located interharmonic at 75 Hz of Case 9, are presented in Figures 2.7.(a) to (e). Both the 50 Hz and the 75 Hz components are extracted with high accuracy regardless of the presence of the supraharmonic component. The extracted features of both 50 and 75 Hz components are observed to have waveform error below 5.2% and RMS error of 0.89.

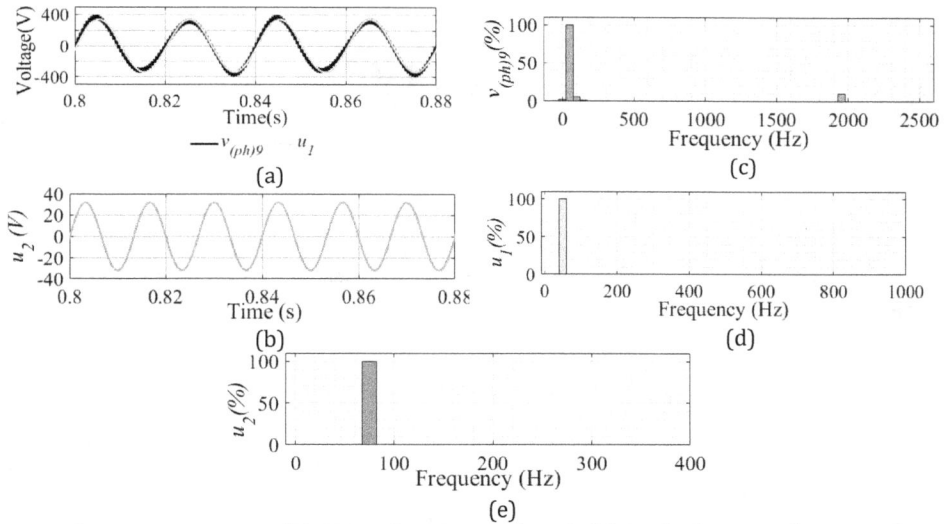

Figure 2.7: Time response of VMD synchronizer in Case 9: (a) Applied $V_{(ph)9}$ and Extracted $u_1$, (b) Extracted $u_2$ and (c) to (e) Frequency spectrum of Applied $V_{(ph)9}$, Extracted $u_1$ and Extracted $u_2$.

The response of the VMD synchronizer for random transient events in IDM are discussed next. The normal and magnified views of applied input voltage $v_{(ph)9}$ with a magnitude surge of 10% and the extracted fundamental frequency component, $u_1$ is depicted in Figure 2.8.(a). The VMD synchronizer has been found extracting the fundamental component successfully despite the presence of transient with zero absolute error in frequency, magnitude and with 0.056% error in phase. Figure 2.8.(b) shows the applied input with multiple zero crossing distortions and the extracted fundamental component from case 11. The extracted fundamental component in Figure 2.8.(b) is seen to be devoid of any zero crossing error.

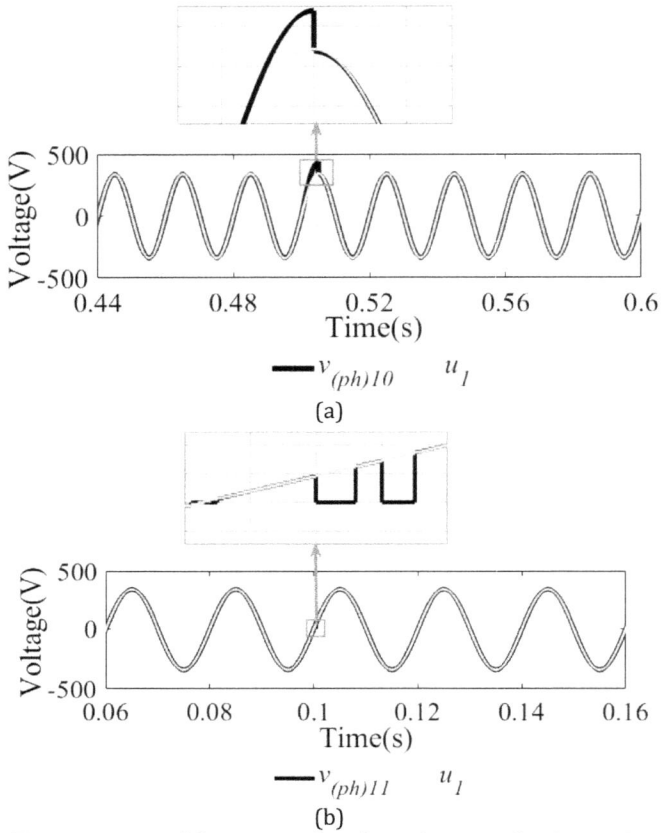

Figure 2.8: Time response of the proposed grid synchronizer for Cases 10 and 11 with magnified view at instant of transition; (a) Applied $v_{(ph)10}$ with a surge at 0.5s to 0.505s and extracted $u_1$ in Case 10 and (b) Applied $v_{(ph)11}$ with zero crossing distortions between 0.1s to 0.1005s and extracted $u_1$ in Case 11.

Magnified views at the instants of transition in Figure 2.8.(a) and (b) show no signature of any transient. These transient events, can be characterized in high frequency range, so are showing up in the high '$k$' bands of VMD modes, thus leaving the extracted fundamental with no sign of the transients.

The real time field data from the distribution feeder has been collected and used as the input signal and the synchronization signal retrieved from the VMD algorithm are presented in Figure 2.9.(a) to (d). The same waveforms are measured using a power quality analyzer and the values are used to cross verify the VMD outputs.

(a)

(b)

(c)

(d)

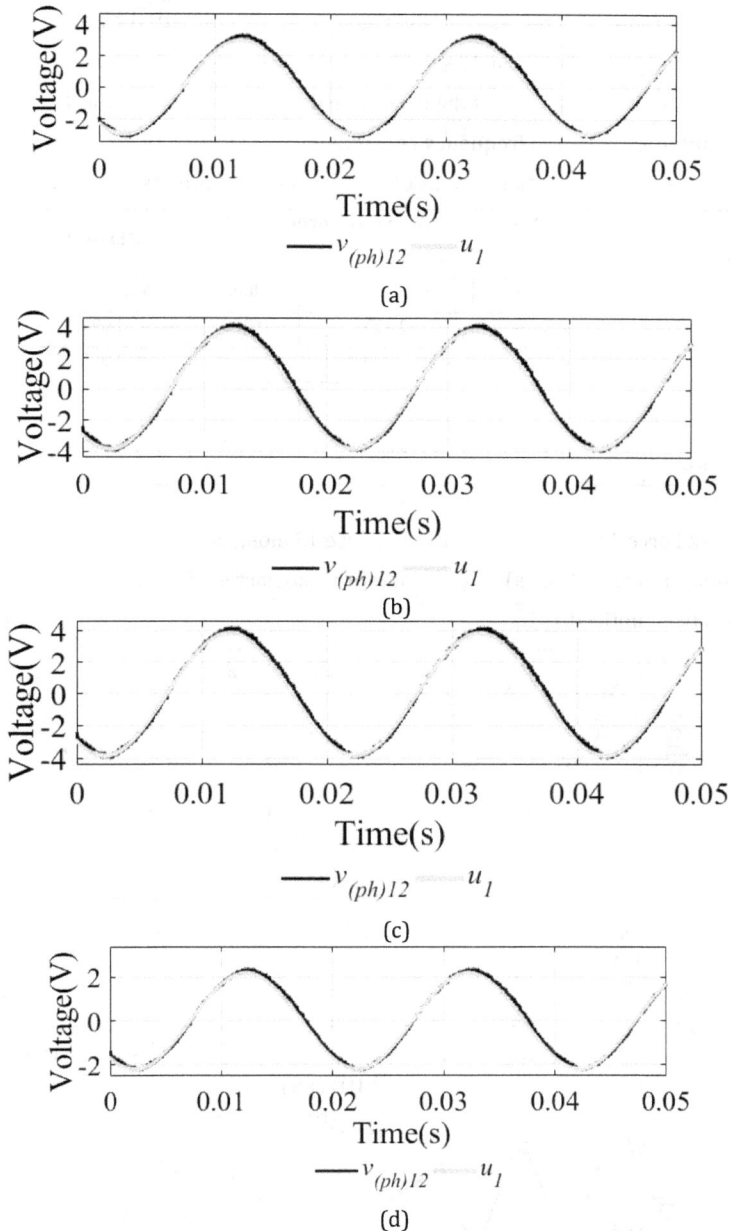

Figure 2.9: Time response of applied $v_{(ph)12}$ and extracted $u_1$ for Case 12, (a). At the incoming feeder of the campus distribution network; (b). At a UPS feeder in the campus, (c). At electrical machines laboratory in the campus; and (d). At power electronics laboratory in the campus.

The results are compared and found to be in good compliance with the measurements by the power quality analyzer (Fluke LM 434) in magnitude, frequency and phase. These results are presented in Table 2.3 which shows the maximum deviation to be as low as 0.37%, 0.05% and 0.2% in magnitude, phase and frequency respectively.

TABLE 2.3 EXTRACTION ERROR COMPUTED FOR REAL TIME DISTRIBUTION DATA

| Location | Absolute Error in waveform (ε) in % | | | RMS Error in sampled data | |
|---|---|---|---|---|---|
| | Frequency | Magnitude | Phase | Magnitude | Phase |
| Machines Lab | -0.2 | 0.353 | 0.055 | 0.699 | 0.018 |
| UPS Feeder | 0.2 | 0.360 | 0.055 | 0.699 | 0.018 |
| Main Feeder | -0.2 | 0.378 | 0.055 | 0.699 | 0.018 |
| Power Electronics Lab | -0.1 | 0.272 | 0.055 | 0.699 | 0.018 |

The IEEE Task Force 1159.2 data [92] used in Case 13 along with extracted features from VMD are presented in Figure 2.10.(a) to (d) wherein a satisfactory fundamental extraction for all data sets has been noticed.

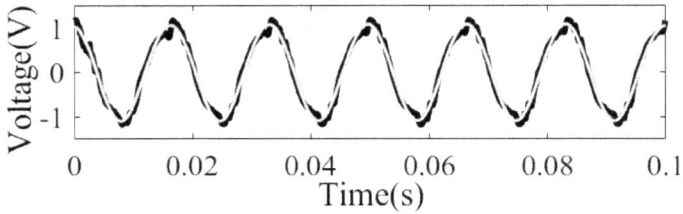

(a)

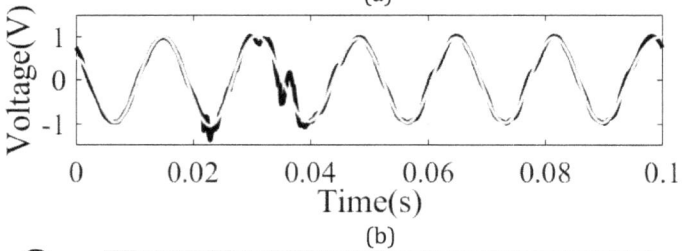

(b)

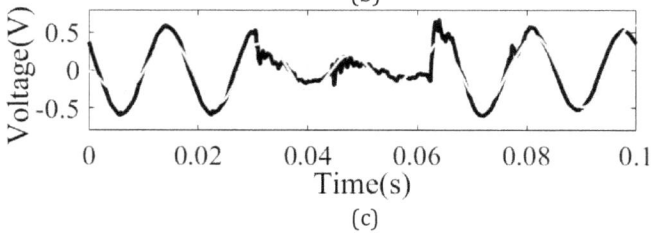

(c)

(d)

$\underline{\quad\quad} v_{(ph)13} \quad\quad u_1$

(e)

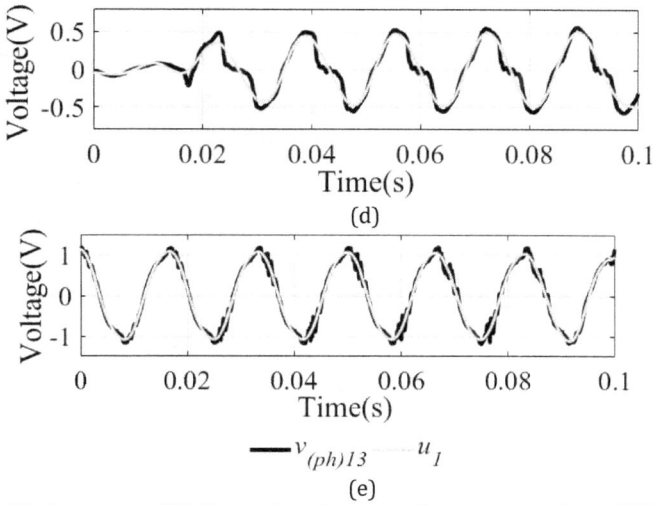

Figure 2.10: Performance of VMD synchronizer with data set taken from IEEE task force 1159.2 for power quality event characterization: (a) to (e)Time response of applied input $v_{(ph)13}$ and extracted $u_1$ for data 1 to 5 respectively.

The error indices were computed based on the signal composition data given by IEEE Task Force corresponding to each case and are presented in Figure 2.11.

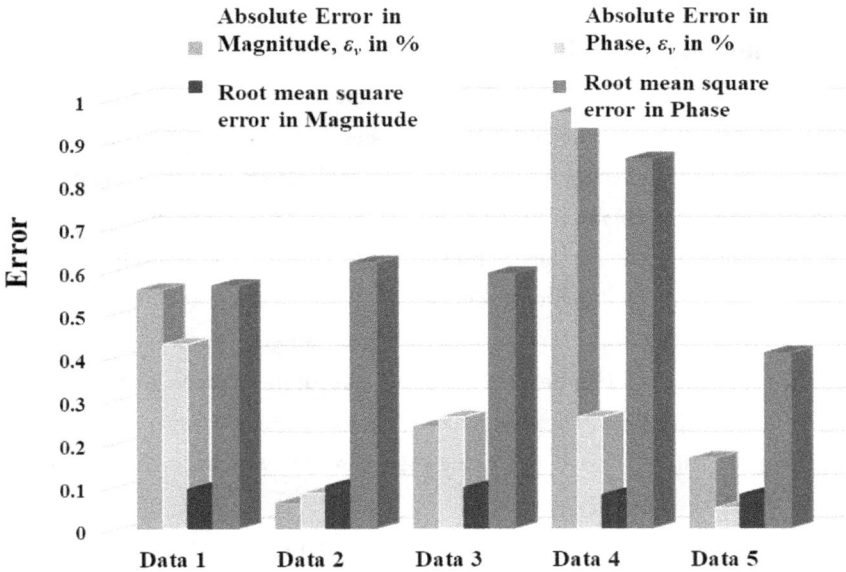

Figure 2.11: Error indices computed for IEEE 1159.2 task force for power quality event characterization data set 1 to 5 for case 13.

- **Summary of results:**

The error indices of all test cases are summarized in Table 2.1 which indicate high accuracy levels in frequency extraction with negligibly small error 'ε' values. The highest error in magnitude at sample levels for VMD is found to be 0.7465, while the cycle level error is zero across all cases, for extracted fundamental component. The maximum error in the phase information of the extracted fundamental is found to be 0.1% with a lowest RMS phase error of 0.0281.

These error indices obtained are much lower in comparison to the PLL based synchronizers [46], [50], [51], [55], [91], [95]–[97]. The VMD synchronizer is also found to be competent to extract the grid voltage fundamental frequency while handling new age multiple power quality issues like supraharmonics, interharmonics, subharmonics, etc. The VMD synchronizer also exhibited a commendable accuracy in tracking deviations in fundamental frequency, magnitude, and phase angle. The frequency tracking of VMD synchronizer can also be further improved by increasing the input signal sampling rates. The VMD synchronizer has been observed to possess an innate immunity to reject random transients. The random transients are high frequency components, thus when decomposed with VMD synchronizer these transients get shifted to higher modes, resulting in an error less fundamental frequency mode.

## 2.5.3 Interdependency of the VMD parameters $k$ and $\alpha$

VMD parameters have substantial influence on the extraction/reconstruction accuracy especially when working with higher number of modes. But, in synchronizer, it is sufficient to segregate the fundamental component from the rest of the frequency components. However, there exists an inverse relationship between the accuracy of the extracted fundamental component with the presence of higher frequencies in MCS and the allowable bandwidth at each mode. This is due to the interdependency of VMD parameters $k$ and $\alpha$ on the mode segregation and capture of minor principle frequency deviation. Hence a sensitivity analysis is conducted to recognize this interdependency so as to advocate a systematic way of designing the VMD parameters for synchronizer as it has to work with unknown spectral bands.

The following input signal, $f$, is applied to the VMD algorithm with the objective of extracting the fundamental frequency component.

$$f = \sqrt{2} \times 240 sin(2\pi 50t) + \sqrt{2} \times 48\ sin\ (2\pi 250t) + \sqrt{2} \times 24\ sin\ (2\pi 350t)$$
$$+\sqrt{2} \times 12\ sin\ (2\pi 550t) + \sqrt{2} \times 12\ sin\ (2\pi 750t)$$

In the developed VMD algorithm, $k$ is assigned a value of 5 which is equal to the number of frequencies in '$f$' and $\alpha$ is varied between 800 to 80,000. The errors in the extracted fundamental component observed for this variation are presented in Figure 2.12 in which $\alpha$ = 8000 gives the lowest error. Choice of any lower or higher value of $\alpha$ outside this range resulted in incomplete extraction beyond certain frequency band.

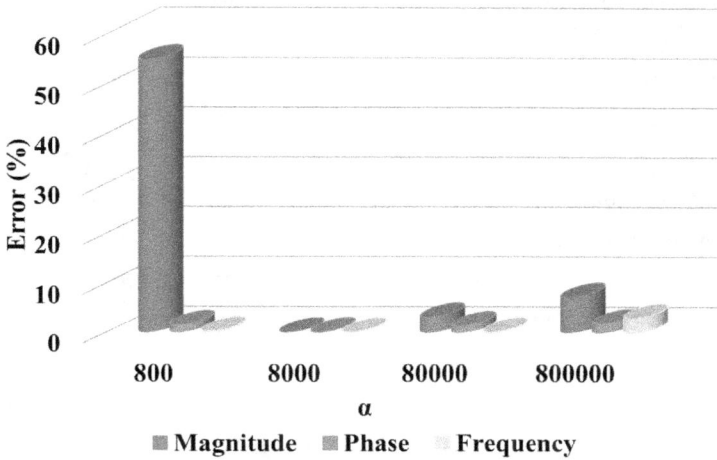

Figure 2.12 Influence of $\alpha$ on extraction accuracy with $k$ =5 (Case 1a)

Conversely, $k$ is varied from 0 to 12, keeping $\alpha$ at the identified optimum value of 8000 for the same test case 1a. This range of $k$ is chosen in order to have a wide variation in the mode. The result of this investigation is presented in Figure 2.13 which shows that the extraction error is the least when the $k$ value matches with the number of the harmonic frequencies in the input signal.

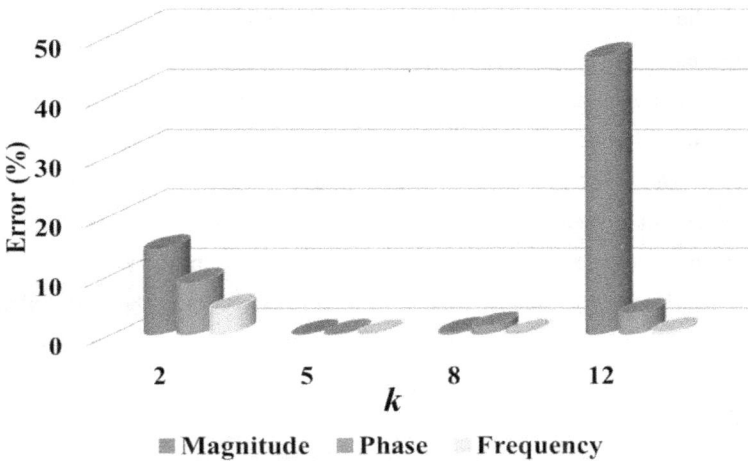

Figure 2.13 Influence of k on extraction accuracy with $\alpha$ = 8000. (Case 1a)

41

But the error is significantly higher for lower $k$ values. When deployed in evolving inverter dominated smart grids, the archived grid data can guide a right selection of $k$ value based on the most prevalent frequencies and $\alpha$ can be selected based on the frequency regulation history.

The inferences from the sensitivity analysis have been used in the design of the VMD synchronizer. The design constraints applied for selection of all the VMD parameters are presented below.

- **Choice of $k$:** For complete and accurate decomposition, $k$ must be equal to the number of constituent frequencies present in the spectral band of MCS [68], [69]. But in IDMs, the grid voltage is anticipated to be a nonstationary MCS with unknown spectral composition If $k$ is smaller than the total number of constituent frequencies then the feature extraction accuracy may be reduced due to mode mixing [68]. Further, the fundamental frequency is expected in one of the lower modes, it will not suffer from the inaccuracy due to mode mixing. Thus, a $k$ value higher than the average number of constituent frequencies is deemed to be ideal for the proposed VMD synchronizer.

- **Choice of $\alpha$:** The prime objective of grid synchronizer is a fine separation of the fundamental frequency; so, the Weiner filter function is made narrow and centered around the grid frequency with a small sideband so that minor frequency deviations will be captured without fail [68], [88]. Higher values of $\alpha$ result in a narrow filter bandwidth which eventually does a finer frequency separation, but it will fail to capture minor deviations of the principal frequency. In contrast, lower values of $\alpha$ widen the filter bandwidth, thereby facilitating the capture of a wide band variation of the fundamental frequency, but results in inaccurate separation [68]. However, extremely narrow bandwidths may not recognize the minor fundamental frequency shifts which are inherently present in electric grids. So, a mid-range value of $\alpha$ is found appropriate for the proposed application.

- **Choice of $\mu$:** *This* is the Lagrangian multiplier whose value ranges from 0 to 1. Selecting a particular value of $\mu$ determines the priority of constraint function in equation (2.7). In VMD, the objective function is mode segregation, and the constraint function is reconstruction. The purpose of VMD synchronizer is accurate feature extraction at the mode corresponding to the fundamental frequency which demands mode segregation to be prioritized to the reconstruction [68], [69]. So, $\mu$ is designed to be zero for the VMD synchronizer.

- **Choice of $\tau$ and $n$:** In VMD, the value of central frequency of a mode, $\omega$, is iteratively computed as per equation (2.9). The difference between two consecutive center frequencies,

42

$\omega_k{}^n$ and $\omega_k{}^{n+1}$ will decide when the iteration can be terminated for a particular mode's central frequency. If the difference between $\omega_k{}^n$ and $\omega_k{}^{n+1}$ is less than or equal to $\tau$, the optimization can be terminated for that particular mode. If the number of iterations reach $n$, then the iterations stop even if the difference between $\omega_k{}^n$ and $\omega_k{}^{n+1}$ is not near $\tau$. With larger value of $\tau$ or smaller value of $n$, the VMD algorithm can achieve faster mode segregation, but with reduced accuracy. A very small value of $\tau$ with a large value of $n$, improves the accuracy and yields finer mode segregation, but with increased computation time. Thus, a lower value of $\tau$ with a mid-value of $n$ is identified to give an accurate mode segregation along with a favorable computation time, which best suits the demands of a synchronizer.

- **Input window size:** This refers to the minimum number of cycles of the MCS required by the VMD synchronizer for extraction of fundamental frequency component. High window size allows a reliable decomposition process, but yields poor dynamic response, which is not acceptable in synchronizers. On the contrary, low window size can improve the dynamic response, yet it can cause incomplete decomposition due to insufficient signal length. Generally, the window size is chosen as the full cycle period of the lowest frequency component to be extracted [68], [69][68], [69], but VMD theory has proved to reliably decompose any MCS with a period as low as quarter cycle of the frequency to be extracted. Therefore, a minimum window size equal to quarter cycle period of the fundamental frequency is chosen with a moving window in this work.

The proposed VMD synchronizer has been designed and tested in simulation and in hardware experiments.

# 2.6 Comparative analysis between EWT and VMD for grid synchronization

## 2.6.1 EWT as a grid synchronizer

As explained in section *2.3.1*, yet another decomposition algorithm, EWT, also possess the characteristics suitable to meet the challenges of a grid synchronizer. [66]-[67], [70]-[71], [82]-[83]. An attempt is made in this research to design and develop a EWT synchronizer and to compare its performance with that of the VMD synchronizer.

## 2.6.2 Design considerations for EWT synchronizer

A sampled input of $f(t)$ is applied to the EWT synchronizer, following which all the high amplitude components in the frequency spectrum are picked out and grouped as $F(\omega)$. The central frequencies of elements in $F(\omega)$ are then grouped as equivalent frequency vector, $\omega_n$, where $n=0,1,2,....N$ and $N$ is the total number of modes [66]-[67], [70]-[71], [82]- [83]. The Fourier spectrum, ranging from $[0,\pi]$ is then segmented into $N$ parts and each segment is identified as $\Lambda_i$, where $\Lambda_i =[\; \omega_{n-1}\; ;\; \omega_n]$. This normalizes the equivalent frequencies between the boundaries, $\omega_0=0$ and $\omega_N=\pi$, in each segment. So, the mid frequency, $\omega_i$, between the boundaries represents the center of two consecutive maximum points while the Fourier segments become $[0,\; \omega_1]$, $[\omega_1,\; \omega_2]$,.........., $[\omega_{N-1},\; \pi]$. Such a spectrum segmentation essentially identifies the possible dominant frequency constituents in a MCS [66]-[67], [70]-[71], [82]- [83].

An additional segment called a transition phase, $T_n$, is then added along with the $\omega_i$ to form small side bands that accommodate a small deviation in the central frequency. This attribute of EWT can be utilized to design the EWT synchronizer to accommodate the fundamental grid frequency deviations. The width of $T_n$ is defined as $2\tau_n$, where $\tau$ is a design criterion [66]-[67], [70]-[71], [82]- [83].

A set of wavelet functions are then derived empirically to form adaptive filters for mode segregation. The *Mayer's* wavelet unit scaling function, *Mayer's* wavelet function and *Mayer's* auxiliary function, $\beta(y)$, given by equations (2.19), (2.20) and (2.21) respectively are reported to have accurate time and frequency resolution for non-stationary signal decomposition [66]- [67], [70]-[71], [82]- [83]. Hence, these are chosen as the wavelet functions for the proposed EWT synchronizer.

$$\hat{\emptyset}_n(\omega) = \left\{ 1\; if\; |\omega| \le \omega_n - \tau_n \; cos\; cos \left[\frac{\pi}{2}\beta\left(\frac{1}{2\tau_n}(|\omega| - \omega_n + \tau_n)\right)\right] \; if\; \omega_n - \tau_n \le \right.$$
$$\left. |\omega| \le \omega_n + \tau_n \; 0\; otherwise \right\} \quad .........(2.14)$$

$$\hat{\varepsilon}_n(\omega) = \left\{ 1\; if\; \omega_n + \tau_n \le |\omega| \le (1-\gamma)\omega_{n+1} - \tau_{n+1}\; cos\; cos \left[\frac{\pi}{2}\beta\left(\frac{1}{2\tau_{n+1}}(|\omega| - \right.\right.\right.$$
$$\left. \omega_{n+1}+\tau_{n+1})\right)\right]\; if\, \omega_{n+1} - \tau_{n+1} \le |\omega| \le \omega_{n+1}+\tau_{n+1}\; sin\; sin \left[\frac{\pi}{2}\beta\left(\frac{1}{2\tau_n}(|\omega| - \omega_n + \right.\right.$$
$$\left.\left. \tau_n)\right)\right]\; if\, \omega_n - \tau_n \le |\omega| \le \omega_n + \tau_n\; 0\; otherwise \right\} \quad .........(2.15)$$

$$\beta(y) = y^4(35 - 84y + 70y^2 - 20y^3) \quad .........(2.16)$$

The inner product of $F(\omega)$ with $\hat{\emptyset}_n(\omega)$ computes the approximate wavelet coefficients while the inner product of $F(\omega)$ with $\hat{\varepsilon}_n(\omega)$ computes the detailed wavelet coefficients [66]-[67], [70]-

[71], [82]- [83]. Since $\tau_n$ is appended with central frequency of each mode, the equations (2.14) and (2.15) can be modified as,

$$\hat{\phi}_n(\omega) = \left\{ 1 \ if \ |\omega| \leq (1-\gamma)\omega_n \ cos \ cos \left[\frac{\pi}{2}\beta\left(\frac{1}{2\gamma\omega_n}(|\omega|-(1-\gamma)\omega_n)\right)\right] \ if (1-\gamma)\omega_{n+1} \leq |\omega| \leq \right.$$
$$\left. (1+\gamma)\omega_{n+1} \ if (1-\gamma)\omega_n \leq |\omega| \leq (1+\gamma)\omega_n \ 0 \ otherwise \right\} \quad .......... (2.17)$$

$$\hat{\varepsilon}_n(\omega) = \left\{ 1 \ if \ (1+\gamma)\omega_n \leq |\omega| \leq (1-\gamma)\omega_{n+1} \ cos \ cos \left[\frac{\pi}{2}\beta\left(\frac{1}{2\gamma\omega_n}(|\omega|- \right.\right.$$
$$(1-\gamma)\omega_{n+1})\right] \ if (1-\gamma)\omega_{n+1} \leq |\omega| \leq (1+\gamma)\omega_{n+1} \ sin \ sin \left[\frac{\pi}{2}\beta\left(\frac{1}{2\gamma\omega_n}(|\omega|- \right.\right.$$
$$(1-\gamma)\omega_n)\right] \ if (1-\gamma)\omega_n \leq |\omega| \leq (1+\gamma)\omega_n \ 0 \ otherwise \right\} \quad ........(2.18)$$

where, $\tau_n = j*\omega_n$, and $j=0<j<1$. Once these adaptive wavelet filers are employed, they form a boundary across the dominant modes. This boundary then moves empirically to segregate the signal [66]-[67], [70]-[71], [82]- [83].

In EWT, the Fourier component of the applied grid voltage is segregated into $N$ modes which can be defined by the user. For synchronizer application, $N = 6$. Similar to VMD, each mode has a main central frequency with a small sideband. The boundary support for first mode, with $\omega_0$ = 0, is defined by default. Therefore, $N+1$ output modes are retrieved for segregation of the MCS into N modes [66]-[67], [70]-[71], [82]- [83]. The first mode with frequency of $\omega_0$ is usually discarded unless the input has any DC component. The 50 Hz or 60 Hz fundamental frequency component will be mostly available as 2nd mode.

### 2.6.3 Testing and validation of EWT synchronizer

The designed EWT synchronizer is validated in MATLAB/Simulink. The present-day grid voltage signal with four odd harmonics ranging from the 5th onwards till 13th is used as test input, $v_a$, and is made to decompose all the constituent frequency components. $v_a$ is shown in Figure 2.13 is described as,

$$v_a = \sqrt{2} \times 240 \ sin \ (2\pi50t) + \sqrt{2} \times 48 \ sin \ (2\pi250t) + \sqrt{2} \times 24 \ sin \ (2\pi350t) + \sqrt{2} \times 12$$
$$sin \ (2\pi550t) + \sqrt{2} \times 12 \ sin \ (2\pi750t) \quad ........(2.18)$$

The absolute errors in magnitude, phase and frequency are used to assess the accuracy of extraction.

The response of both the EWT and VMD synchronizers for extraction of fundamental grid voltage signal from $v_a$ is presented in Figure 2.14.

45

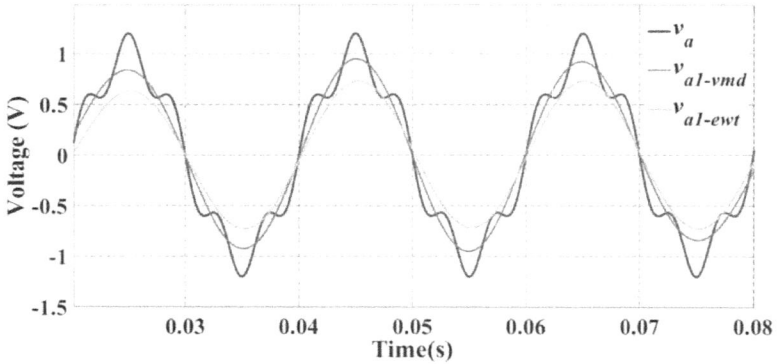

Figure 2.14: Time responses of VMD synchronizer, $v_{a1\text{-}vmd}$, and EWT synchronizer, $v_{a1\text{-}ewt}$ for contemporary grid signal, $v_a$.

The fundamental component extracted by VMD, $v_{a1\text{-}vmd}$ and EWT, $v_{a1\text{-}ewt}$ are observed to have zero frequency error, while the magnitude and phase error is slightly higher in EWT compared to VMD. However, the overall performance of EWT synchronizer is still found to be much superior to most of the advanced PLLs. Thus, both EWT and VMD synchronizers exhibit comparable extraction accuracies while working with grid voltages with single power quality issue or minor frequency deviation.

The test signals from IEEE Task force 1159.2 [92] are used to test the EWT synchronizer and the results are compared with that of VMD synchronizer.

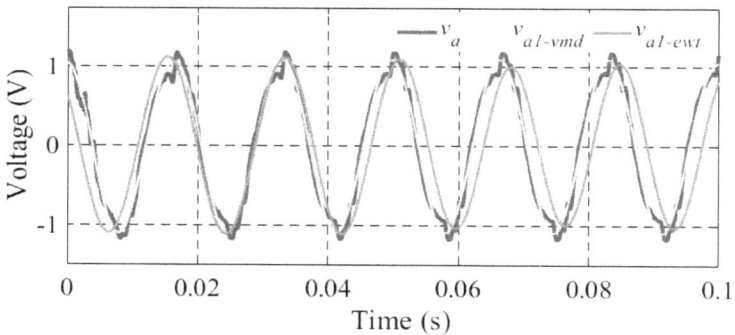

(a)

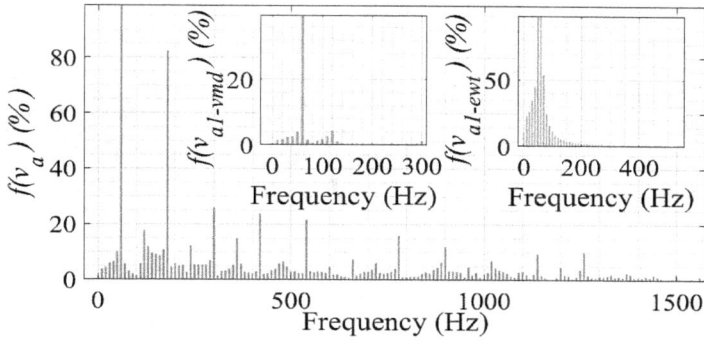

(b)

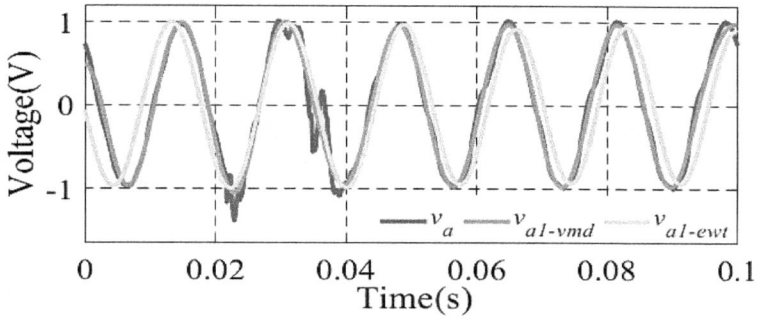

(c)

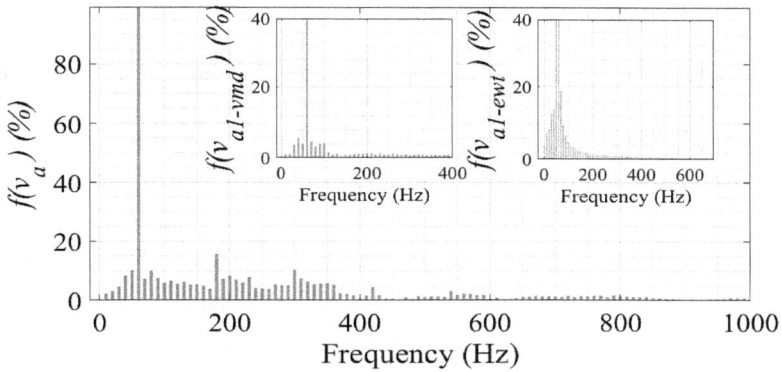

(d)

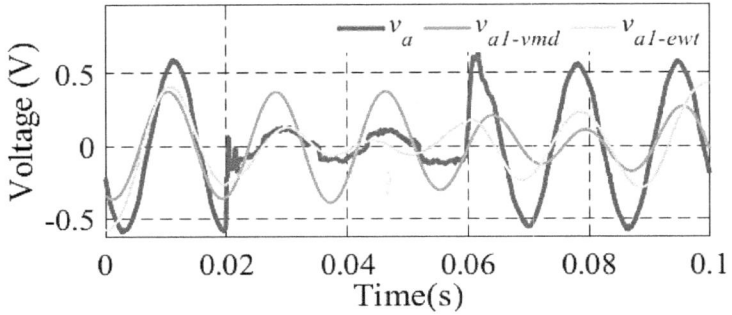

(e)

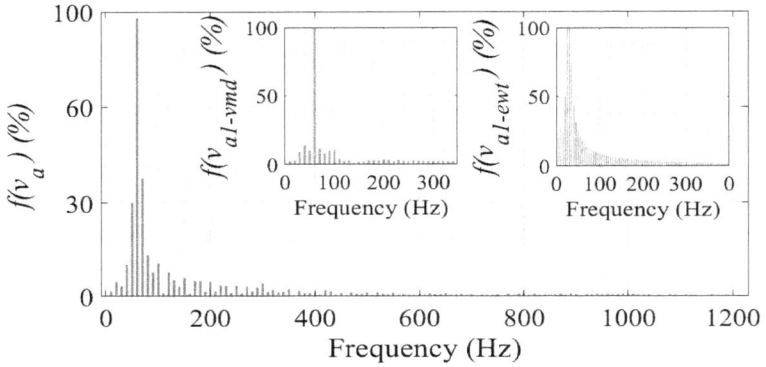

(f)

Figure 2.15: Time responses of VMD and EWT synchronizers for IEEE Task force 1152.9 data sets 1 to 3: (a), (c) and (e). $v_a$, and $v_{a1\text{-}vmd}$ and $v_{a1\text{-}ewt}$, for data sets 1,2 and 3 respectively;. (b),(d) and (f). Frequency spectra of $v_a$, and $v_{a1\text{-}vmd}$ and $v_{a1\text{-}ewt}$, for data sets 1,2 and 3 respectively.

TABLE 2.4 PERFORMANCE COMPARISON BETWEEN VMD SYNCHRONIZER AND EWT SYNCHRONIZER

| Signal input | VMD synchronizer | | | EWT synchronizer | | |
|---|---|---|---|---|---|---|
| | Absolute error in waveform | | | Absolute error in waveform | | |
| | Magnitude $\varepsilon_v$ (%) | Phase $\varepsilon_p$ (%) | Frequency $\varepsilon_f$ (%) | Magnitude $\varepsilon_v$ (%) | Phase $\varepsilon_p$ (%) | Frequency $\varepsilon_f$ (%) |
| Contemporary grid signal | 0 | 0 | 0 | 0.34 | 0.01 | 0 |
| Data 1 | 0.56 | 0.1 | 0.01 | 1.14 | 2.14 | 2.84 |
| Data 2 | 0.06 | 0.1 | 0.07 | 1.2 | 2.45 | 9.35 |
| Data 3 | 0.24 | 0.097 | 0.14 | 12.54 | 14.24 | 5.32 |

48

The results corresponding to IEEE task force input signals are presented in Figure 2.15.(a) to (f) and the respective error indices are summarized in Table 2.4. In Figure 2.15.(a), the extracted component, $v_{a1-ewt}$, is observed to have magnitude, phase and frequency errors of 1.14%, 2.14% and 2.84% respectively corresponding to data set 1. The higher frequency error of 2.84% in extracted $v_{a1-ewt}$ seen in Figure 2.15.(b) is indicative of mixing in lower frequency modes. The extraction accuracy of the EWT synchronizer is found to be lower than that of VMD synchronizer for data set 2 as seen in Figure 2.15. (c). The frequency spectrum of, $v_{a1-ewt}$ corresponding to data set 2 in Figure 2.15.(d) comprises many closely located low frequencies rather than a single central frequency. Whereas the VMD synchronizer is able to achieve extraction of fundamental frequency component with 0.07% error for data set 2.

The input data and the extracted fundamental signals for set 3 of $v_a$, containing multiple frequency deviations, harmonic combinations, and magnitude fluctuations are shown in Figure 2.15.(e), in the results corresponding to this data set are given in Table 2.4. From the frequency spectrum of Figure 2.15.(f), mode mixing is observed in both VMD and EWT synchronizers. But VMD has mode mixing much less than that of EWT synchronizer, indicating that the VMD synchronizer is more competent in handling the multiple power quality issues and non-linearities which are much anticipated in IDM.

In EWT, the IMF is essentially the Mayer's wavelet whose bandwidth is then adaptively varied using the empirical mode decomposition. Though Mayer's wavelet has been proven to have better extraction accuracy for non-linear signals, it is observed to be reduced when used for processing signals with unknown spectral composition. On the other hand, aptly tuning the optimization parameters in a VMD synchronizer makes it competent in decomposing nonstationary and nonlinear signals.

Therefore, a hardware prototype of the VMD synchronizer is then developed in the laboratory and tested for its real time performance, as reported in the following section.

## 2.7 Real time testing of proposed VMD synchronizer

A primary validation of the proposed VMD synchronizer is attempted through a scaled down real time laboratory testing. The testing is executed for a set of emulated signals with a fundamental frequency of 20Hz and having various power quality issues; single event as well as multiple events have been tested separately.

## 2.7.1 Test bench description

The test bench considered with workflow is depicted in Figure 2.16. The test signals are generated via emulations using MATLAB/Simulink models. The developed simulations for generation of the test signals are run for infinite time in order to mimic a real time operation. All the signals are developed with a fixed sampling rate of 50μs.

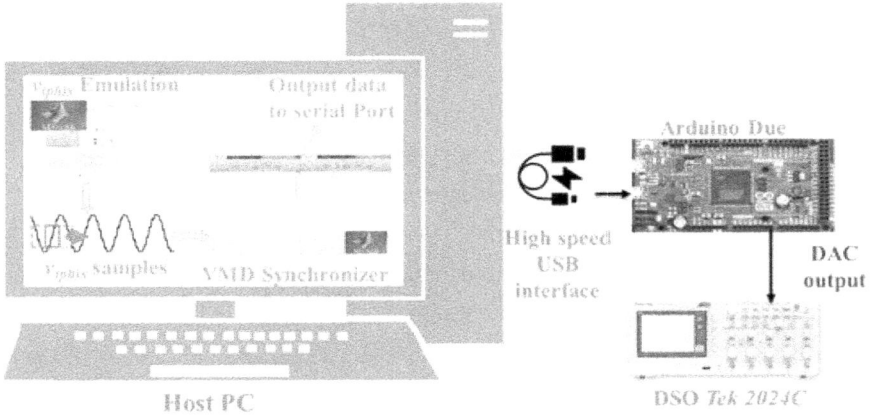

Figure 2.16: Laboratory test bench arrangement with workflow for real time validation of proposed synchronizer

The input, $v_{(ph)x}$ generated from simulation is then processed by accumulating samples for a quarter cycle window and then passed on to the VMD synchronizer function. The VMD synchronizer is coded in the MATLAB as a function with $v_{(ph)x}$ as input argument and $u_1$ extracted as output. Both $v_{(ph)x}$ and $u_1$ are then communicated over a high-speed serial port to UART/USB interface to DAC of Arduino Due board. The input and the output analog signals are visualized in a DSO *Tek 2024C*.

## 2.7.2 Test results, analysis and discussion

The performance of the proposed VMD synchronizer has been validated for 10 test cases on the test bench shown in Figure 2.16. The test cases are summarized in Table 2.5. In all cases, the test signals have been emulated to have a fundamental frequency of 20Hz owing to the communication speed restrictions of Arduino UART. For ease of processing, the test signals are represented in p.u with a base of 1V RMS and +1.6V DC offset.

TABLE 2.5 TEST CASES FOR REAL TIME VALIDATION OF VMD SYNCHRONIZER

| Case No. | Condition/power quality event | Description of $v_{(ph)x}$ |
|---|---|---|
| 1 | Low Order Harmonics | $v_{(ph)1} = \sqrt{2} \times 240\ sin\ sin\ (\ 2\pi20t) + \sqrt{2} \times 48\ sin\ sin\ (\ 2\pi100t) + \sqrt{2} \times 24$ $sin\ sin\ (\ 2\pi140t) + \sqrt{2} \times 12\ sin\ sin\ (\ 2\pi220t) + \sqrt{2} \times 12$ $sin\ sin\ (\ 2\pi260t)$ <br> Contemporary Grid signal with lower order harmonics from 5th to 13th |
| 2 | Suprharmonics | $v_{(ph)2} = \sqrt{2} \times 240\ sin\ sin\ (2\pi20t)\ + \sqrt{2} \times 48\ sin\ sin\ (2\pi100t)\ + \sqrt{2} \times 24$ $sin\ sin\ (2\pi780t)$ <br> Fundamental grid voltage overriding with 5th order lower harmonics and 39th order supraharmonics |
| 3 | Voltage sag | Fundamental grid voltage of 240V RMS overriding with 5th harmonic is subjected to a sag of 10% of the nominal value, at the instant of 0.5s for a period of 0.15 s |
| 4 | Voltage swell | Fundamental grid voltage of 240V RMS overriding with 5th harmonic is subjected to a swell of 10% of the nominal value, at the instant of 0.5s for a period of 0.15 s |
| 5 | Voltage unbalance | Fundamental grid voltage of 240V RMS overriding with 5th harmonic is subjected to a magnitude imbalance of ± 10% introduced as a voltage rise in phase 'a' at the instant of 0.5s for a period of 0.1s, followed by a voltage fall. |
| 6 | Interharmonics | Fundamental grid voltage of 240V RMS overriding with 5th harmonic is subjected a 10% interharmonics at 30Hz and a 10% supraharmonics of 780 Hz, expressed as, <br> $v_{(ph)9} = \sqrt{2} \times 240\ sin\ sin\ (\ 2\pi20t) + \sqrt{2} \times 24\ sin\ sin\ (\ 2\pi30t) + \sqrt{2} \times 24$ $sin\ sin\ (\ 2\pi780t)$ |
| 7 | Voltage surge | Fundamental grid voltage of 240V RMS overriding with 5th harmonic is subjected to a voltage transient of 10% above the nominal value, introduced to the input voltage at the instant of 0.5s for a period of 0.005s |
| 8 | Frequency jump | Fundamental grid voltage of 240V RMS overriding with 5th harmonic is subjected to a frequency deviation of ± 0.5 Hz from its nominal value is introduced at the instant of 0.5s for a period of 0.15s |
| 9 | Phase jump (lag) | Fundamental grid voltage of 240V RMS overriding with 5th harmonic is subjected to a phase angle deviation of -10° lag, at the instant of 0.5s for a period of 0.05 s |
| 10 | Phase jump (lead) | Fundamental grid voltage of 240V RMS overriding with 5th harmonic is subjected to a phase angle deviation of +10° lead, at the instant of 0.5s for a period of 0.05 s |

Figures 2.17 to 2.19 depict the real time responses of VMD synchronizer for test cases 1 to 10 mentioned in Table 2.5. The steady state accuracy of the proposed VMD synchronizer is seen to comply with the simulation results. The VMD synchronizer extracts the fundamental frequency component, $u_1$, without any error in magnitude, phase and frequency when tested with grid voltage signal having lower order odd harmonics as seen in Figure 2.17.(a). The high extraction accuracy of the VMD synchronizer in processing signals with supraharmonics is evident from Figure 2.17.(b). The extracted fundamental component, $u_1$, has been observed to have an absolute error of 0.01%, 0.05% and 0.07% in frequency, magnitude and angle respectively.

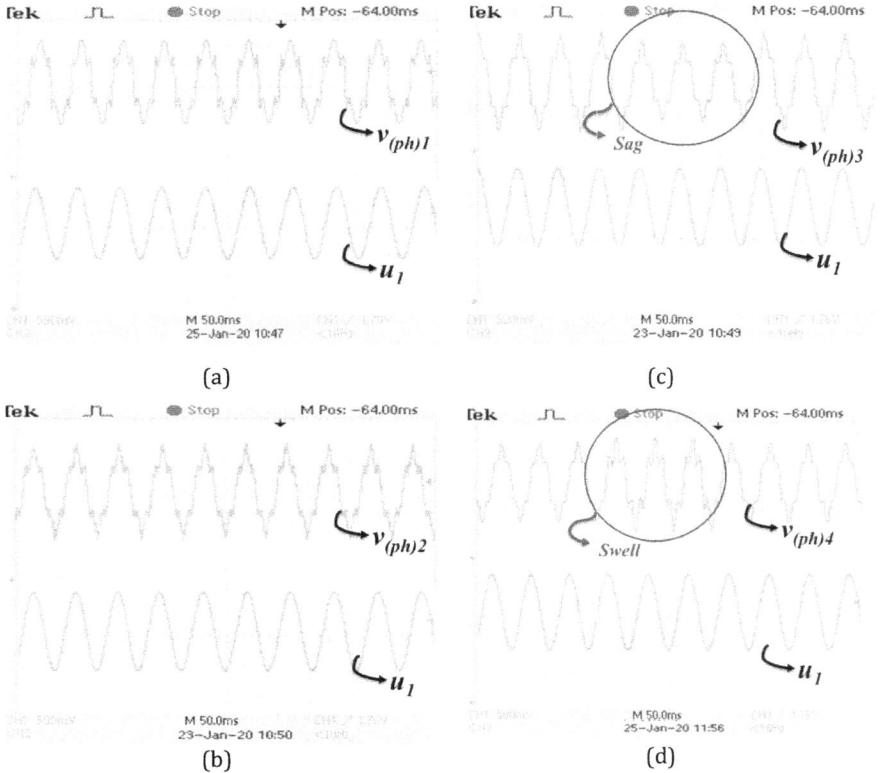

Figure 2.17: Emulated grid voltage '$v_{(ph)}$' and VMD synchronizer output '$u_1$': (a). test case 1, low order harmonics and (b) test case 2 , supraharmonics, (c). test case 3, voltage sag and (d). test case 4, voltage swell respectively.

The VMD synchronizer is found to track the transient voltage sag and swell as depicted in Figure 2.17. (c) and (d) effectively. The VMD synchronizer is also found capable to detect magnitude deviations during an imbalance in one of the phases as seen from Figure 2.18.(a). The extraction accuracy of the VMD synchronizer to decompose closely located fundamental frequency and interharmonic frequency has been tested and the result is given in Figure 2.18.(b); the absolute errors are as low as 0.01%, 0.041% and 0.057% in frequency, magnitude and phase respectively. Figure 2.18.(c) depicts the result when the VMD synchronizer was subjected to a magnitude surge; it validates the synchronizer's immunity against such transient. There is no transient present in $u_1$ in this case. Thus, even under transient events, the signal processing alternative is found to perform with high accuracy and exhibit good dynamic response.

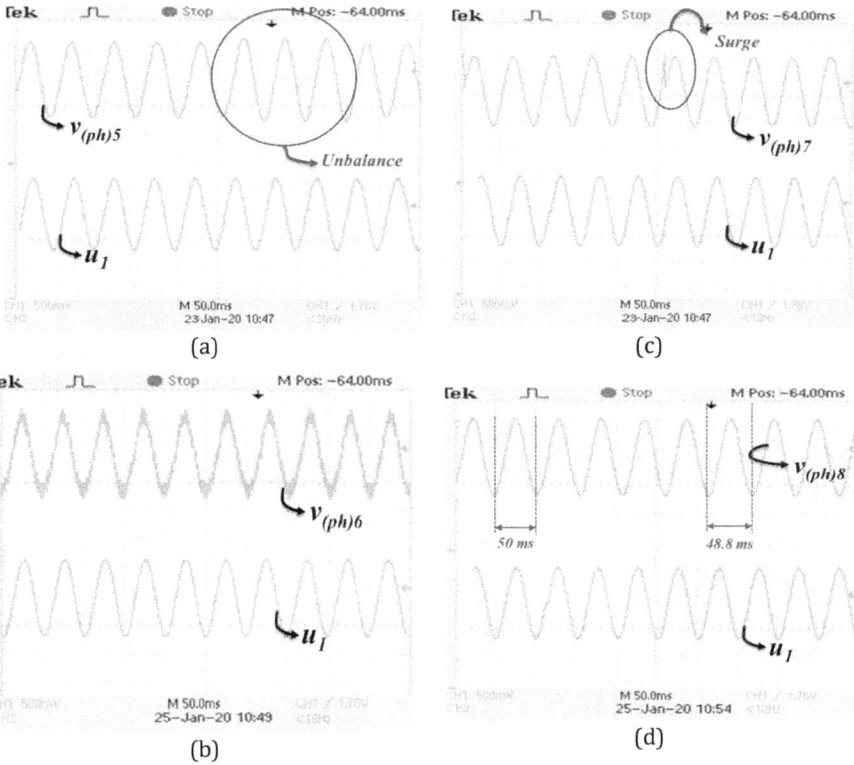

Figure 2.18: Emulated grid voltage '$v_{(ph)}$' and VMD synchronizer output '$u_1$': (a). test case 5, imbalance, (b) test case 6, voltage surge, (c). test case 7, interharmonics and (d) test case 8, frequency jump respectively.

The input signal with a frequency jump of 0.5Hz is presented in Figure 2.18.(d). The synchronizer is observed to follow the frequency change in the input signal and reflects it in the extracted fundamental right from the instant the frequency jump is introduced. The frequency error is observed to be 0.41 % along with 0.04% magnitude error and 0.07% phase angle error.

The VMD synchronizer is found to track deviations in phase angle of voltage signal as seen in Figures 2.19.(a) and (b) with absolute errors of 0.01%, 0.047% and 0.08% in frequency, magnitude and phase respectively.

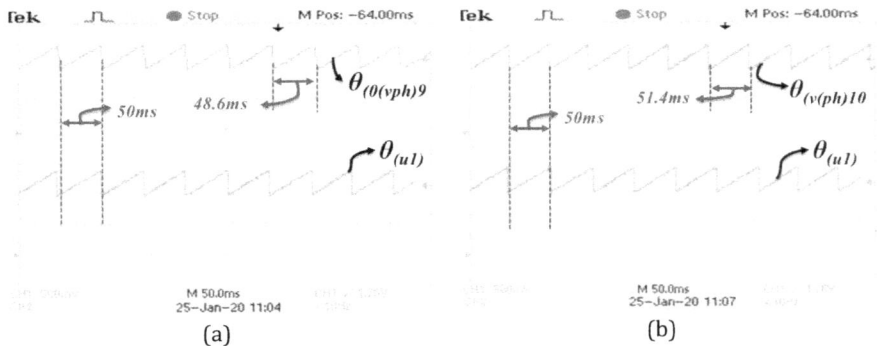

Figure 2.19: Emulated grid voltage '$v_{(ph)}$' and VMD synchronizer output '$u_1$': (a) and (b) Grid voltage phase $\theta_{(v(ph))}$ and synchronizer output phase $\theta_{(u1)}$ in case 9 and 10 respectively.

TABLE 2.6 ERROR INDICES FOR REAL TIME VALIDATION OF VMD SYNCHRONIZER

| Test case | Absolute error in % | | |
|---|---|---|---|
| | Frequency | Magnitude | Phase |
| | ($\varepsilon_f$) | ($\varepsilon_v$) | ($\varepsilon_\theta$) |
| Lower order odd harmonics | 0.01 | 0.04 | 0.08 |
| Supraharmonics | 0.01 | 0.05 | 0.07 |
| Voltage sag and voltage swell | 0.02 | 0.045 | 0.074 |
| Unbalance | 0.01 | 0.041 | 0.057 |
| Phase angle jump (lag) | 0.01 | 0.045 | 0.08 |
| Phase angle jump (lead) | 0.01 | 0.047 | 0.07 |
| Frequency jump | 0.41 | 0.04 | 0.07 |
| Interharmonics | 0.04 | 0.05 | 0.41 |
| Voltage surge | 0.04 | 0.048 | 0.41 |

## 2.8 Comparison of VMD synchronizer with the state-of-the-art synchronizers

The performance of the VMD synchronizer has been compared with that of PLL based synchronizers as well as with other MCS decomposition techniques in decomposing non-stationary and non-linear voltage signal. The results of the comparative studies have been summarized below. The comparison assumes a scale given as follows:

- Very Good: Absolute error between 0-1.5%
- Good: Absolute error between 1.5-2%
- Partially Good: Absolute error between 2-2.5%
- Below Average: Absolute error between 2.5-3.5%
- Average: Absolute error between 3.5-4.5%
- Poor: Absolute error between greater than 4.5%

54

## 2.8.1 Performance of PLL based synchronization methods

A comparison of features and attributes of the various PLL structures, compiled from published literature, is summarized in Table 2.7 [43]–[48], [50], [91], [95]–[101].

TABLE 2.7 COMPARISON OF STATE-OF-THE-ART SYNCHRONIZATION METHODS (ADVANCED PLL TECHNIQUES)

| Features/aspects | | dq | αβ | ddsrf | DSOGI | dαβ | FPD dαβ & Adaptive FPD dαβ | MAF | EMAF | EPMAF (type 1 and 2) & αβEPMAF | MSHDC | DNαβ |
|---|---|---|---|---|---|---|---|---|---|---|---|---|
| Minimum signal length | | 100 samples/s | | | | | | | | | | |
| Nonstationary signal dynamics | Small band deviations | Good | | | | | | | | | | |
| | Wide Band Variations | Poor | Poor | Poor | Good | Average | Average | Below Average | Average | Average | Average | Average |
| Accuracy of Extracted Fundamental Under | Magnitude Deviation | Poor | Average | | | | | | | | | |
| | Phase Deviation | Good | | | | | | | | | | |
| | Frequency Deviation | Poor | Poor | Poor | Poor | Poor | Good | Average | Average | Average | Average | Average |
| | Unbalance | Poor | Poor | Poor | Good | Good | Good | Good | Average | Average | Average | Average |
| | DC Offset | Poor | Poor | Poor | Poor | Good | Good | Good | Average | Poor | Poor | Poor |
| | Off Nominal Frequency | Good | | | | | Poor | Poor | Good but Partially Good for Type 2 | Good | Good | Good |
| | Harmonics | Poor | | | | | Good | Good | Good | Good | Good | Good |
| | Inter and Intra Harmonics | Average | | | | | Partially Good | | Partially Good but Better for Type 2 | Poor | Poor | Poor |
| Immunity | Notches | Average | Average | Average | Average | Average | Average | Average | Average | Average | Average | Average |
| | Impulsive Transients | Average | Average | Average | Average | Average | Average | Average | Average | Average | Average | Average |
| | Oscillatory Transients | Average | Average | Average | Average | Average | Average | Average | Average | Average | Average | Average |
| | Fault Detection | Average | Average | Average | Good | Good | Good | Good | Good | Good | Good | Good |
| | Computational complexity | Very low | Low | Low | Low | High | Very low | Low | Low | Very high | Very high | Very high |
| Real time application logistics | Robustness | Average | Average | Average | Average | Average | Average but very slow | Average but slightly slower | Average but slightly slower | Good and Faster than other PLL | Good and Faster than other PLL | Good and Faster than other PLL |
| | Reliability | Average | Average | Average | Average | Average | Average | Average | Average | Good | Good | Good |
| | Response to Arbitrary Signal | Poor | Poor | Poor | Poor | Poor | Poor | Poor | Poor | Poor | Poor | Poor |

The performance of the proposed VMD synchronizer has been observed to be competent and better than other state-of-the-art PLLs available in literature.

## 2.8.2 Comparison of proposed VMD synchronizer with other signal decomposition algorithms

The extraction features and attributes of the proposed VMD synchronizer have been compared with those of other signal decomposition techniques in Table 2.8 [49], [52], [55]–[58], [60], [62], [65], [66], [68], [71]–[74], [80], [82]–[84], [86], [89], [90], [102]–[111]. It tells clearly that VMD is superior to all other techniques in this art of fundamental frequency extraction.

TABLE 2.8 COMPARISON OF STATE-OF-THE-ART SYNCHRONIZATION METHODS (SIGNAL PROCESSING ALGORITHMS)

| Features/aspects | | Frequency domain methods | | | Time domain methods | | Time- frequency methods | | | | |
|---|---|---|---|---|---|---|---|---|---|---|---|
| | | FT | DFT | FFT | Root Mean Square | EMD | STFT | ST | WT (DWT and WPT) | EWT | VMD |
| Minimum signal length | | 20ms | 100 samples/s | | 20ms | 5ms | Depends on the window width chosen | | Depends on mother wavelet function & decomposition level | 5ms | |
| Time frequency localization | | No | | | | | Yes | | | | |
| Non stationary signal dynamics | Small band deviations | Poor | Better than fft | Better than ft | Na | Good | Good | Better than WT but depends on window size | Good for known signal with matched mother wavelet function and decomposition level | Good (but work depends on some prior assumptions) | Very good |
| | Wide Band Deviations | Poor | NA | Good | Good | Better than WT but depends on window size | Good for known signal with matched mother wavelet function and decomposition level | Good (but work depends on some prior assumptions) | Very Good | NA | Good |
| Accuracy of Extracted Fundamental Under | Magnitude Deviation | Poor | Poor | Poor | Poor | Poor | Poor | Poor | Average | Average | Good |
| | Phase Deviation | Poor | Poor | Poor | Poor | Good | Average /Poor due to fixed window size | Average | Good for known signal with matched mother wavelet function and decomposition level | Good | Very Good (except for higher harmonics with proper choice of number of modes) |

| | | | | | | | | | | |
|---|---|---|---|---|---|---|---|---|---|---|
| **Frequen cy Deviatio n** | Poor | Poor | Poor | Poor | Good | Average /Poor due to fixed window size | Average | Good for known signal with matched mother wavelet function and decomp osition level | Avera ge | Very Good (exce pt for highe r harm onics with prope r choice of numb er of mode s) |
| **Unbalan ce** | Poor | Poor | Poor | Poor | Good | Average /Poor due to fixed window size | Average | Good for known signal with matched mother wavelet function and decomp osition level | Good | Very Good (exce pt for highe r harm onics with prope r choice of numb er of mode s) |
| **DC Offset** | Poor | Poor | Poor | Poor | Good | Average /Poor due to fixed window size | Average | Good for known signal with matched mother wavelet function and decomp osition level | Good | Very Good (exce pt for highe r harm onics with prope r choice of numb er of mode s) |
| **Off Nomina l Frequen cy** | Poor | Poor | Poor | Poor | Average | Average /Poor due to fixed window size | Average | Good for known signal with matched mother wavelet function and decomp osition level | Good | Very Good (exce pt for highe r harm onics with prope r choice of numb er of mode s) |
| **Harmon ics** | Poor | Average | Good | Poor | Good | Average /Poor due to fixed window size | Good | Good for known signal with matched mother wavelet function and decomp osition level | Good | Very Good (exce pt for highe r harm onics with prope r choice of numb er of mode s) |

| | | | | | | | | | | |
|---|---|---|---|---|---|---|---|---|---|---|
| **Im mu nity** | **Inter and Intra Harmon ics** | Poor | Poor | Poor | Poor | Good | Average /Poor due to fixed window size | Average | Good for known signal with matched mother wavelet function and decomp osition level | Avera ge | Very Good (exce pt for highe r harm onics with prope r choice of numb er of mode s) |
| | **Notches** | Poor | Good | Avera ge | Poor | Good | Average /Poor due to fixed window size | Average | Good for known signal with matched mother wavelet function and decomp osition level | Avera ge | Very Good (exce pt for highe r harm onics with prope r choice of numb er of mode s) |
| | **Impulsi ve Transie nts** | Poor | Good | Avera ge | Poor | Good | Average /Poor due to fixed window size | Good | Good for known signal with matched mother wavelet function and decomp osition level | Good | Very Good (exce pt for highe r harm onics with prope r choice of numb er of mode s) |
| | **Oscillat ory Transie nts** | Poor | Good | Avera ge | Poor | Average | Average /Poor due to fixed window size | Good | Poor for noise | Avera ge | Very Good (exce pt for highe r harm onics with prope r choice of numb er of mode s) |
| | **Fault Detectio n** | Poor | Poor | Poor | Poor | Good | Average /Poor due to fixed window size | Good | Good | Good | Very Good (exce pt for highe r harm onics with prope r choice of numb er of mode s) |

| | | | | | | | | | | | |
|---|---|---|---|---|---|---|---|---|---|---|---|
| | **Computational complexity** | Simple | Simple but depends on implementation methods | | Simple | Complex | Slightly complex than FT, DFT and FFT | Complex | | | |
| **Real time application logistics** | **Robustness** | Good for Stationary Signal | | | | Good | Average | Average | Average | Average | Very Good |
| | **Reliability** | Good for Stationary Signal | | | | Good | Average | Average | Average | Average | Very Good |
| | **Response to Arbitrary Signal** | Poor | Poor | Poor | Poor | Poor as filter banks choice depends on input | Poor | Poor | Poor | Poor | Very Good |

### 2.8.3 Discussion on comparative study

The salient merits of the VMD synchronizer upon other signal processing techniques as well as on the advanced PLL algorithms employed for synchronization application can be summarized as:

(i). Capability to detect, track and adapt to even a narrow band of grid frequency variation,

(ii). Significantly low settling time in capturing frequency deviations.

(iii). Immunity to multiple power quality events, random noise and high frequency transients, and, voltage sags and swells

(iii). Veracity and adaptability to tackle inputs even with unknown spectral bands.

These exclusive features together with the other merits deduced from this primary investigation further support a claim that VMD has an edge over EWT as a synchronizer for the grid connected converters to work in the emerging IDM.

Overall, the proposed synchronizer is seen to have competent performance with good extraction accuracy, no mode mixing at low frequencies, extraction under fundamental frequency deviation and better dynamics in comparison to almost all the advanced PLLs.

## 2.9  Summary

Design, development, and testing of VMD algorithm as a grid synchronizer has been executed. The VMD and EWT have been identified from a range of signal decomposition algorithms and designed to meet the challenges as a grid synchronizer. The results have been presented and the key findings discussed in this chapter. Major highlights of this chapter are presented below:

- A feasibility study has been conducted to identify a suitable MCS decomposition technique which could serve as a grid synchronizer for highly dynamic IDM.
- VMD and EWT techniques have been identified as two potential candidates to be designed as signal decomposed grid synchronizer.
- VMD is proved to be better than EWT as a synchronizer with less than 0.6% error in magnitude, phase and frequency tracking respectively.
- The VMD grid synchronizer exhibited shorter extraction time in the range of 1.4 ms to 2 ms with high extraction accuracy while handling a variety of fluctuating signatures of grid voltage.
- The VMD parameters are designed to extract only the fundamental frequency. The higher order frequency components extracted with the same design showed up small errors, which can be reduced by appropriate redesign when required in other applications like harmonic computations.
- The observations under various test conditions ascertain that the signal decomposed synchronizer possesses immunity against most of the power quality disturbances and events.
- The VMD synchronizer exhibited a superior dynamic performance under various transient events like sag, swell, phase and frequency jumps, surges, etc.; the transition time under these events has been observed to be short with the tracking accuracy as high as 99% during and after the transient events.
- The VMD synchronizer has been found to reject almost all the random transient events with no sign of these in the extracted signal.

# CHAPTER 3

# Development of Model Predictive Control for Grid Inverter

## 3.1 Introduction

The modelling and design perspectives necessary to develop a model predictive controlled grid-tied inverter for IDM has been presented in this chapter. It is followed by a sensitivity analysis to study the influence of sampling frequency of MPC performance in grid-tied inverters. Further, the design, development and validation of a MPC based grid-tied inverter with bidirectional power flow are presented.

## 3.2 Inverter power control with nonlinear controller

Grid-tied inverters are the fundamental units in IDMs, to control the active and reactive powers delivered and are also responsible for maintaining stability and reliability of the network [15]. Control of three phase grid connected inverters involve voltage or current control to achieve the desired targets [22], [40], [98]. Basic idea of current control involves in inverter switching so as to ensure inverter delivered current tracking the reference current with minimum error in amplitude, shape and phase.

From the literature survey of chapter 1, a class of implicit modulation technique has been identified as a potential candidate for controlling grid tied inverter systems. These class of controllers do not involve a modulator to obtain actuation unlike voltage control techniques thus extending a flexibility in the control. But such nonlinear direct control suffers from issues like variable switching frequency that affects inverter stability, increased filter size requirement to block lowest order harmonic, etc. [20], [114]–[117].

However, nonlinear controllers are robust systems with better dynamic response, instantaneous waveform correction, inherent peak current protection, overload rejection, better load dynamics, reduced delay, etc. [22], [30], [36], [40], [98], [118]–[122]. These characteristics

Model Predictive Control (MPC), an offshoot of nonlinear predictive control, is an approach capable of achieving additional control targets besides reference tracking [38], [117], [123]–[136]. MPC is identified as a candidate which has the potential to eliminate the major drawbacks of hysteresis based control, sliding mode control, etc., thus proving to be an attractive choice for the future inverter control [38], [117], [123]–[136]. This research work identified MPC as the choice for grid tied inverter control and utilize it for IDM environment. An effort to analyze the performance of MPC based grid tied inverter for its tracking capability, harmonic profile and total inverter power loss and the same is reported here.

## 3.3. Modelling and design of model predictive controlled grid-tied inverter

MPC is first applied for a grid tied inverter and its characteristics and performance are studied for predictive model parameter variations. This can help in designing MPC to suit the control demands of IDMs with targeted performances. The accuracy of MPC highly dependent on the accuracy of the load model and its execution in discrete time domain. This section briefs the development of the load model and its implementation for a grid tied inverter.

A model predictive controlled grid-tied inverter system is shown in Figure 3.1. This inverter is assumed to be working in a microgrid environment with a power reference commanded by a local microgrid controller.

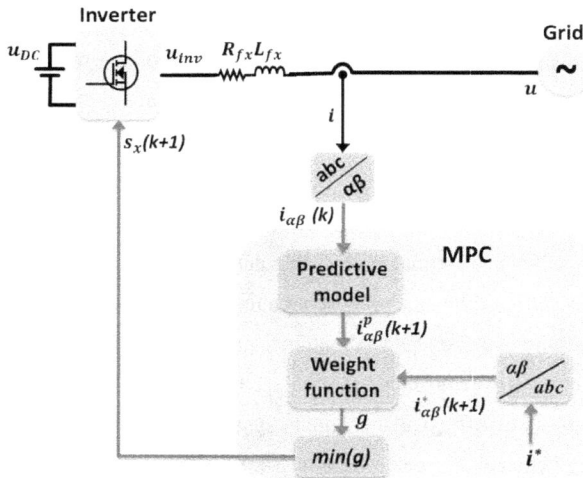

Figure 3.1: Model predictive controlled grid-tied inverter system

The system comprises a three phase, two level inverter, an inductive filter and the model predictive current controller unit. The current reference for the MPC is computed from the power reference as demanded by the local microgrid controller. The input to the inverter is assumed to be provided by the RE sources available in the IDM.

### 3.3.1 Modelling and design

MPC is a sample-based technique in which a mathematical model of the system under consideration is used to iteratively predict control variables and optimize it to meet the control constraints. MPC can handle system nonlinearities and control constraints simultaneously while meeting the main control tasks. This is in contrast to the classical linear controllers which often try to neglect or simplify such system nonlinearities. Selection of iteration size and optimization tuning are two important design procedures to ensure proper control. Hence an extensive analysis of this emerging control technique is of vital significance for selection of appropriate digital controllers for its real time implementation. Further, the minimum operating frequency of the inverter and its influence on the filter and heat sink sizing can also be fixed based on this analysis.

In the present work the design of MPC scheme for grid-tied inverter presented as three stages, which are:

(i). Model development

(ii). Weight function formulation

(iii). Online optimization and actuation

**(i). Model Development**: A mathematical load equation of the grid-tied inverter system has been used to formulate the prediction model of MPC. The input DC voltage, $u_{dc}$, of the grid-tied inverter could be from a solar photovoltaic system or grid storage system like battery. Applying KVL at the inverter output of Figure 3.1, the resulting load current dynamics can be expressed as,

$$u_{inv} = L_{fx}\frac{di_x}{dt} + R_{fx}i_x + u_x \quad ; \{x = a, b, c\} \dots\dots\dots (3.1)$$

where $u_{inv}$ is inverter leg voltage, $i_x$ is the per phase inverter current, $u_x$, is per phase grid voltage, and, $L_{fx}$, and, $R_{fx}$, are the per phase equivalent filter inductance and resistance respectively. The suffix $x$ indicated the phase.

Alternatively, $u_{inv}$, can be expressed using the inverter switching states and DC input voltage as,

$$u_{inv} = u_{dc}(s_a + as_b + a^2s_c) \dots\dots\dots(3.2)$$

where, $a = e^{j2\pi/3}$, and, $s_a$, $s_b$ and $s_c$, represent status, either 0 or 1 of the top switches of the inverter legs. Thus, a three-phase inverter with six switches will have eight possible voltage vectors represented as, $u_0$ to $u_7$, from 000 to 111.

Converting equation (3.2) into its discrete time model by applying Euler's approximation with a sampling frequency, $F_s$, gives,

$$\frac{di_x}{dt} \approx \frac{i_x(k+1)-i_x(k)}{T_s} \text{ where } T_s = {}^1\!/_{F_s} \dots\dots\dots(3.3)$$

where $k$ represents the present sampling instance and $k+1$ is the subsequent sampling instance. Substituting the derivative term in equation (3.1) with equation (3.3) gives,

$$u_{inv}(k) = L_{fx}\left(\frac{i_x(k+1)-i_x(k)}{T_s}\right) + R_{fx}i_x(k) + u_x(k) \quad .........(3.4)$$

Solving for the current, $i_x(k+1)$, by rearranging equation (3.4) results in,

$$i_x(k + 1) = \left(1 - \frac{R_{fx}T_s}{L_{fx}}\right)i_x(k) + \frac{T_s}{L_{fx}}\left(u_{inv}(k) - u_x(k)\right) \quad .........(3.5)$$

In equation (3.5), $R_{fx}$ and $L_{fx}$, are the system parameters, $i_x(k)$, is the measured inverter current and, $u_x(k)$, is the measured grid voltage. The measured current and grid voltages will be supplied by the outer control loop. By substituting the eight possible values of inverter voltage, $u_{inv}(k)$, as $u_0$ to $u_7$ in equation (3.5), computation of all possible values of the current, $i_x(k+1)$, is feasible at the $k^{th}$ sample itself. These future current values are called current predictions given as,

$$i^p(k + 1) = \left(1 - \frac{R_{fx}T_s}{L_{fx}}\right)i_x(k) + \frac{T_s}{L_{fx}}\left(u_z - u_x(k)\right), \ z = 0,1,2,....7 \quad .........(3.6)$$

Equation (3.6) is the discrete time predictive model of a conventional grid-tied inverter system to be used for current tracking by MPC.

**(ii). Weight function formulation**: MPC is a nonlinear control that employs prediction and optimization features to achieve multiple control targets which are weighted as per the application's demand. Essentially, it is a receding horizon [20] optimization problem that uses a discrete time or state space model of the system to predict the future behavior of control variables until a horizon in time.

The optimal actuation to reflect the necessary control action is provided based on minimizing a weight function, which represents the desired target to be attained by the control as,

$$g = f_m + \lambda f_n \quad .........(3.7)$$

where, $f_m$ and $f_n$, represent the primary and secondary targets respectively and, $\lambda$, is a weighting parameter.

The weight function, $g$, in equation (3.7) is a combination of various sub-functions catering to multiple control requirements. Some of these sub-functions represent the hard constraint or primary target and others form soft constraint or secondary targets. Some common primary targets chosen in grid tied inverter control include current, power, torque, etc. [20], [39]. In several applications the converters are expected to accomplish secondary targets such as switching frequency reduction [20], [126], [131], harmonic profile improvement [20], [126], [131], reactive power control [20], [126], [131], flux control, capacitor voltage balancing for multilevel inverters etc. [20], [126], [131]. In MPC, both the primary functions and the

64

secondary functions are achieved with a single weight function. The strict optimization of $f_m$ results in accurate reference tracking, while the secondary targets are prioritized based on the requirement of the auxiliary performance improvement by using the weighting parameter.

Weight function can be formulated either with a single control objective or with multiple control objectives. The weight function can also be developed in any reference frame which does not alter the tracking capabilities of MPC. For basic power tracking in grid-tied inverter in this thesis, the weight function, $g$, is developed to track a current reference in stationary reference frame as,

$$g = |i_\alpha^*(k+1) - i_\alpha^P(k+1)| + \left| i_\beta^*(k+1) - i_\beta^P(k+1) \right| \quad \text{.........(3.8)}$$

where $i_\alpha^*(k+1)$, $i_\beta^*(k+1)$, $i_\alpha^P(k+1)$ and $i_\beta^P(k+1)$, are the stationary reference frame quantities of the reference current, $i_x^*$, and the predicted current, $i^P$, respectively for the $k+1^{th}$ sampling instance. So, during each sampling instance, $g$ is computed as a current error vector with eight values.

**(iii). Online optimization and actuation**: The value of weight function, $g$, in equation (3.8) is computed for every possible inverter voltage vector from $u_0$ to $u_7$. The inverter switching state resulting minimum value of $g$ is selected as the optimal switching state. The gating signals, $s_x$, corresponding to this optimal switching state is applied for the $k+1^{th}$ sampling instance for actuation of the grid-tied inverter.

## 3.3.2 MPC design parameters

The sampling frequency, $F_s$, and the system parameters are the main design variables in the model of the MPC scheme developed in case of grid-tied inverters. The system parameters are $R_{fx}$ and, $L_{fx}$, which can be used directly from the filter specification, while the sampling frequency, $F_s$, is usually left to designers' discretion. Small values of sampling frequency result in fewer computations and cause poor reference tracking performance. On the contrary, a very large value of $F_s$ may result in a large volume of computations, with very short time duration for each burdening the digital processor. Also, this may result in incomplete computations, especially when the digital processor is working with slower clock frequencies and can cause tracking failure. So, selecting an apt value of $F_s$ to match the processor capabilities is important. Having found no guidelines for selection of sampling frequency or comments on its influence on the performance of MPC in literature, a sensitivity analysis on selection of sampling frequency and its influence on the performance of a MPC grid-tied inverter has been carried out.

# 3.4 Sensitivity analysis of sampling frequency variation on performance of MPC

The performance of the three-phase model predictive controlled grid-tied inverter under varying values of sampling frequency has been studied. Four performance indices, viz. tracking accuracy, average inverter switching frequency, current harmonic profile and inverter power loss, have been measured in each case for a comparative assessment.

## 3.4.1 Simulation and analysis

A grid-tied three phase inverter system with MPC scheme described in section *3.4.* is developed in MATLAB/Simulink platform with the specifications given in Table 3.1. Equations (3.6) and (3.8) form the predictive model and weight function, respectively.

TABLE.3.1 System Specifications

| Parameters | Value |
|---|---|
| DC source voltage | 700 V |
| Inverter VA rating | 15 kVA |
| Grid specifications | 3Φ,400V(l-l),50 Hz |
| Filter inductance | 0.001Ω, 10mH |

*Modelling and development of MPC in MATLAB/Simulink:* The grid-tied inverter system under consideration is designed to receive the power references as inputs while the power electronic circuit is developed using inbuilt Simulink library blocks. The proposed MPC is implemented using the 'Embedded MATLAB Function' block in the Simulink model. The prediction system model and weight function are coded in this to mimic the functionality of any real-time controller giving the optimized switching pattern for the grid-tied inverter. The inverter performance index of total harmonic distortion, *THD*, is obtained from the MATLAB inbuilt block of Fast Fourier Transform, FFT, while the average switching frequency and power loss have been computed by customizing the 'User Defined Function' blocks. The proposed control is tested with different conditions to assess the effect of sampling frequency variation on its performance.

66

### 3.4.2 Test Cases for the sensitivity analysis

**Test Case 1**: A three phase current reference, $i^*_x$, of 15.19A (rms) representing a reference power of 14.87kW is applied to MPC block which is working with a sampling frequency of 25 kHz.

**Test Case 2**: Test case 1 is repeated for sampling frequencies of 50 kHz and 100 kHz.

**Test Case 3**: A step change in the reference current, $i^*_x$, is applied at specific time instants as presented in Table 3.2 with a sampling frequency of 50 kHz.

TABLE.3.2 STEP CHANGE IN REFERENCE CURRENT FOR TEST CASE 3

| Time (s) | 0.1 | 0.2 | 0.3 |
|---|---|---|---|
| Power Reference (kW) | 11.57 | 14.87 | 9.47 |
| Change in reference current, $i^*_x$ (as % of 15.19 A) | 10 | 0 | -10 |

**Test Case 4**: A step change of ±10% is introduced to grid voltage, $u_x$ at the mentioned time points as given in Table 3.3 with a sampling frequency of 50 kHz.

TABLE 3.3 STEP CHANGE IN GRID VOLTAGE FOR TEST CASE 4

| Time (s) | 0.1 | 0.2 | 0.3 |
|---|---|---|---|
| Change in grid voltage $u_x$ (as % of 415 V) | 10 | 0 | -10 |

**Test Case 5**: A step change in grid frequency of ±0.5 Hz is introduced at the mentioned time points as given in Table 3.4 with a sampling frequency of 50 kHz.

TABLE 3.4 STEP CHANGE IN GRID FREQUENCY FOR TEST CASE 5

| Time (s) | 0.1 | 0.2 | 0.3 |
|---|---|---|---|
| Change in grid frequency (Hz) | 0.5 | 0 | -0.5 |

The response of the proposed grid-tied inverter for all these test cases are summarized and discussed in the following section.

### 3.4.3 Observations of the sensitivity analysis

The grid-tied inverter performance is quantified using the three indices, viz. tracking accuracy in %, harmonic profile of inverter current in % *THD* and inverter power loss, $P_{loss}$, in W. The tracking accuracy is articulated in terms of magnitude tracking accuracy, $mag_i$, and phase angle tracking accuracy, $ang_i$, both in % as,

$$mag_i = \left(1 - \left(\frac{i^*_{xrms} - i_{xrms}}{i^*_{xrms}}\right)\right) \times 100 \quad ......... (3.9)$$

$$ang_i = \left(1 - \left(\frac{\theta^* - \theta}{360°}\right)\right) \times 100 \quad ........(3.10)$$

where, $i^*_{xrms}$, $i_{xrms}$, $\theta^*$, and, $\theta$, are the rms values and phase angles of the reference current and the inverter delivered current respectively. In MPC, during each sampling interval, the inverter switching state varies depending upon the outcome of the weight function optimization. So, the switching of the inverter is not at a fixed frequency, rather it varies as dictated by the change of state of the inverter due to weight function optimization. Thus, MPC inverter is identified to have a varying switching frequency. Thus, an average value of the inverter switching over a fundamental period is termed as the Average switching frequency, $F_{asw}$, is used to quantify the number of switching changes of the MPC inverter. In this work, the $F_{asw}$ has been computed by manipulation of the switching states of the inverter within a given simulation time.

The total power loss of the inverter with the proposed MPC has been estimated using the fundamental equations [2] and has been found to be as high as 1.01% of the rated inverter power for the average switching frequency of 36.87 kHz. The performance indices are summarized in Table 3.5, which ascertain that an MPC inverter has a commendable reference tracking performance with acceptable efficiency level in grid connected applications.

TABLE 3.5 PERFORMANCE INDICES DEDUCED FROM SENSITIVITY ANALYSIS

| Test Case and condition | $i^*_{xrms}$ (A) | $F_s$ (kHz) | $u_x$ (V) | Grid frequency (Hz) | $i_{xrms}$ (A) | THD (%) | $F_{asw}$ (kHz) | Tracking accuracy (%) | | $P_{loss}$ (W) |
|---|---|---|---|---|---|---|---|---|---|---|
| | | | | | | | | $mag_i$ | $ang_i$ | |
| 1. Track constant current | | 25 | | | 15.08 | 3.73 | 18.7 | 99.28 | 99.83 | 142.2 |
| 2. Track constant current | 15.19 | 50 | | | 15.14 | 1.69 | 36.49 | 99.68 | 99.92 | 144 |
| | | 100 | 400 | | 15.17 | 0.87 | 77.17 | 99.87 | 99.94 | 146.3 |
| 3. Track step change in current | 16.71 | | | 50 | 15.14 | 1.78 | 36.61 | 99.68 | 99.92 | 147.5 |
| | | | | | 16.67 | 1.67 | 36.61 | 99.77 | 99.92 | 147.5 |
| | 13.67 | | | | 13.62 | 2.1 | 36.61 | 99.65 | 99.92 | 147.5 |
| 4. Track with a step change in grid voltage | | | | | 15.14 | 1.78 | 36.7 | 99.68 | 99.92 | 144.2 |
| | | | 440 | | 14.82 | 3.58 | 36.7 | 97.58 | 99.78 | 144.2 |
| | | 50 | 360 | | 15.16 | 1.7 | 36.7 | 99.80 | 99.87 | 144.2 |
| 5. Track with a step change in grid frequency | 15.19 | | | | 15.13 | 1.89 | 38.09 | 99.60 | 99.92 | 149.6 |
| | | | | 50.5 | 15.14 | 1.8 | 38.09 | 99.68 | 99.92 | 149.6 |
| | | | 400 | 49.5 | 15.13 | 1.82 | 36.57 | 99.60 | 99.92 | 149.6 |

## (a). Stead state analysis

The reference current, actual current and other system variables for constant current tracking in test cases 1 and 2 with a sampling frequency of 25 kHz are depicted in Figures 3.2.(a) to (d). The actual current is seen to track the reference current closely with a small error of 0.72%. The grid voltage, $u_x$, and inverter pole voltages, $u_{inv}$, are shown in Figure 3.2.(c) and (d). The magnitude as well as phase tracking accuracy for this case have been observed to be around 99% and are presented in Table 3.5. In Figure 3.2.(e), the trajectory of $i^*_x$, and $i_x$, are seen to exhibit a high degree of agreement with each other except minor deviations, while the current trajectories for 50 kHz and 100 kHz sampling frequencies are presented in Fig. 3.3 (a) & (b).

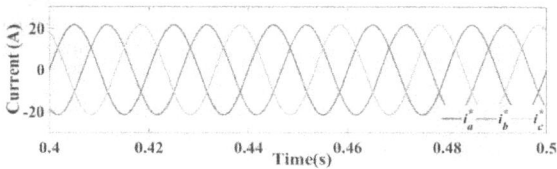

(a)

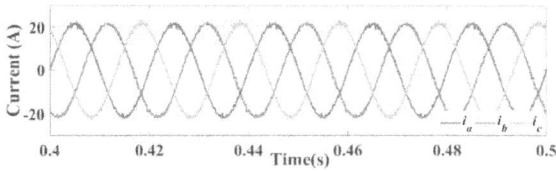

(b)

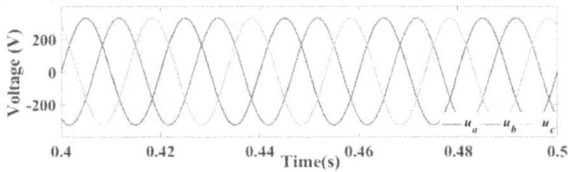

(c)

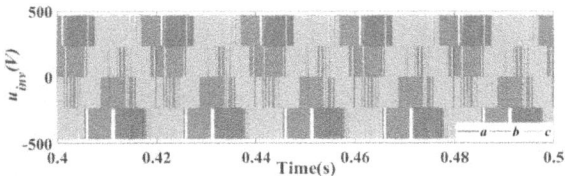

(d)

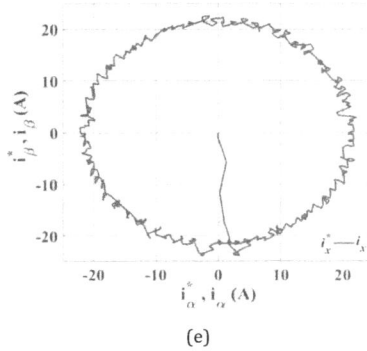

(e)

Figure 3.2: Time Response of proposed control for constant current tracking at 25 kHz sampling frequency: (a) Reference current, $i^*_x$, (b) Actual current, $i_x$, (c) Grid Voltage, $u_x$, (d) Inverter pole voltages, $u_{inv}$, and (e) Current vector trajectory of reference current, $i^*_x$, and actual current, $i_x$, in stationary reference frame.

From Figure 3.3 (a) & (b) it is evident that tracking accuracy increases with increase in sampling frequency. At higher sampling frequency, MPC works at large switching frequencies, resulting in increased accuracy, but at the cost of increased switching loss as seen in Table 3.5. A reduction in current error is observed with increase in sampling frequency from Figure 3. 4.(a) to (c) where, $i_{aerr}$, is the current error in phase a.

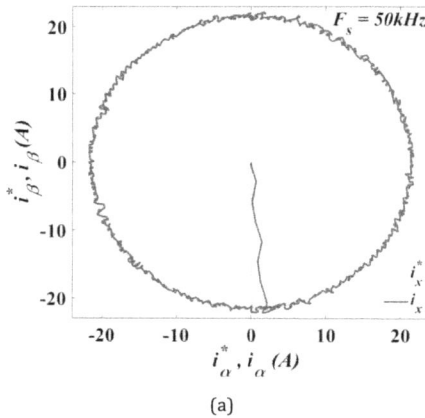

(a)

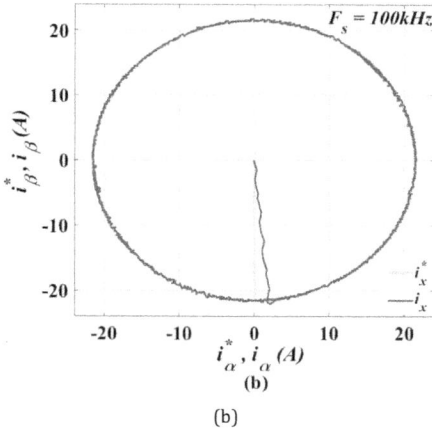

(b)

Figure 3.3:(a) & (b) Current vector trajectory of reference current, $i^*_x$, and actual current, $i_x$, in stationary reference frame coordinates for constant current tracking with sampling frequency of 50kHz and 100kHz respectively.

The harmonic spectrum of inverter current is observed to spread up to half of the sampling frequency. The harmonic profile of the actual current at 50 kHz sampling frequency is shown in Figure 3.4.(d) as a sample case wherein the spectrum is seen to be spread up to 25 kHz.

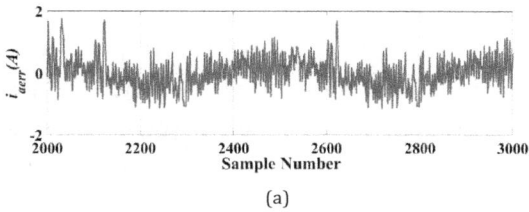

(a)

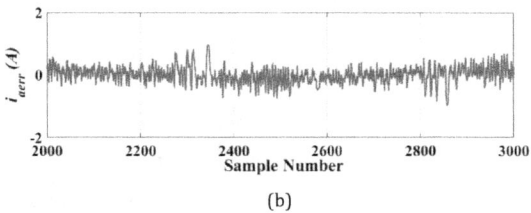

(b)

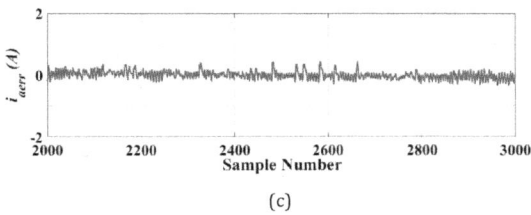

(c)

71

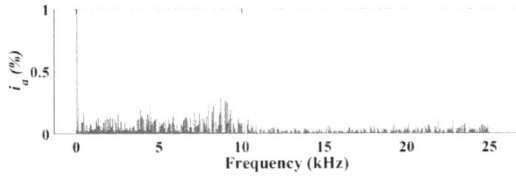

(d)

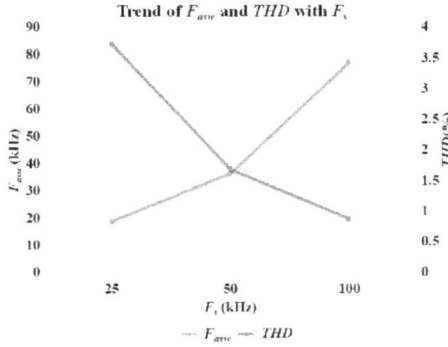

Trend of $F_{asw}$ and $THD$ with $F_s$

$\cdots F_{asw} \quad - \quad THD$

(e)

Figure 3.4:(a) to (c) Instantaneous current error, $i_{aerr}$, for constant current tracking at sampling frequencies of 25, 50 and 100kHz respectively, (d) Current harmonic spectrum for 50 kHz sampling frequency, (e) Trend of $F_{asw}$ and $THD_i$ with varying sampling frequencies.

Figure 3.4.(e) represents the trend of average switching frequency and THD of the inverter current for sampling frequency variations. The average switching frequency is observed to increase and THD is decreased with increase in sampling frequency. Considering both quality of current delivered and reduction of total inverter losses, sampling frequency of 50 kHz is found to be an optimal choice as seen in Figure 3.4. (e). Hence all the remaining test cases are carried out at a sampling frequency of 50 kHz.

**(b). Dynamic Analysis**

Test cases 3 to 5 studied the dynamic analysis of the MPC for a step change each in the reference current amplitude, the grid voltage amplitude and the grid frequency at fixed sampling frequency of 50 kHz. These test cases include the step change in power reference, step change in grid voltage and frequency. A few salient results of these test cases are presented in Figures 3.5 to 3.8.

At 50 kHz sampling frequency, the proposed MPC is found to maintain a tracking accuracy of 99% each in magnitude and phase angle despite a step change in reference current. The tracking dynamics of MPC for step change in current magnitude can be seen in the current

72

vector trajectories of Figure 3.5. The trajectory I belongs to the reference current, $i_x^*$, at steady state with 15.19A rms whereas trajectories II and III correspond to step current changes of ±10%. Here the actual current, $i_x$, is seen to move from trajectory I to II and III in accordance with the change in $i_x^*$, with short transition time and without reduction in tracking accuracy.

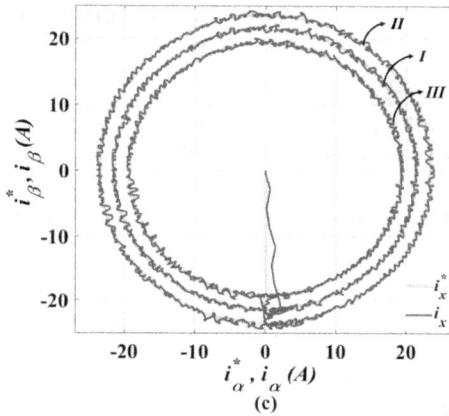

(c)

Figure 3.5: Vector trajectory of reference current, $i_x^*$ Vs actual current, $i_x$; trajectory II for +10% and Trajectory III for -10% change in the reference current.

Figure 3.6 depicts the normal and magnified views of rms value of reference current and actual current for the same test case. The transition time for these 10% step changes are found as 250 μs. This is much less than one sampling interval considering 50 kHz sampling frequency. In test cases 4 and 5, the tracking performance of MPC for step change in grid voltage or grid frequency is examined.

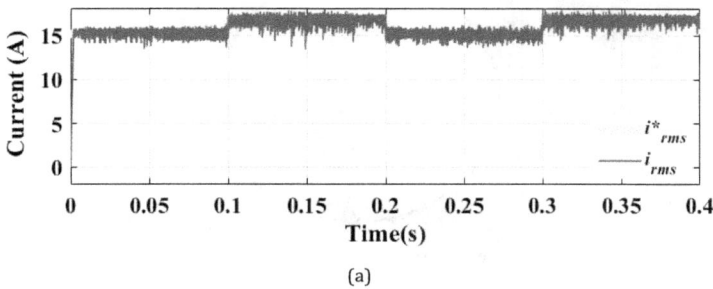

(a)

73

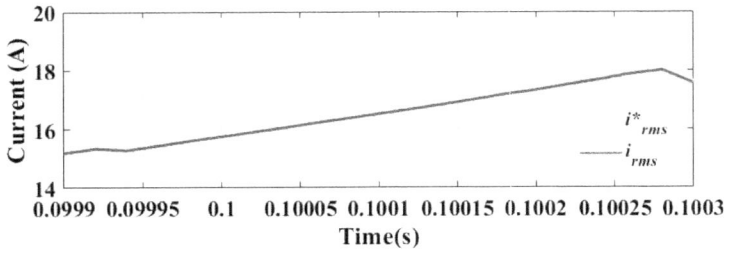

(b)

Figure 3.6: (a) $i^*_x$ and, $i_x$, with ±10% step change at 0.1s and 0.3 s respectively, (b). Magnified views of $i^*_x$ and $i_x$ at an instant of step change.

The current vector trajectory of inverter current in Figure 3.7.(a) is found to trail almost along the trajectory of the reference.

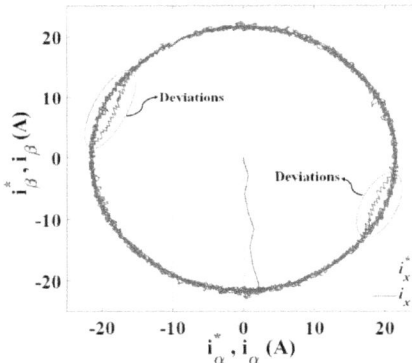

(a)

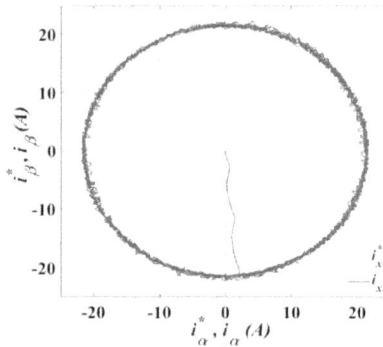

(b)

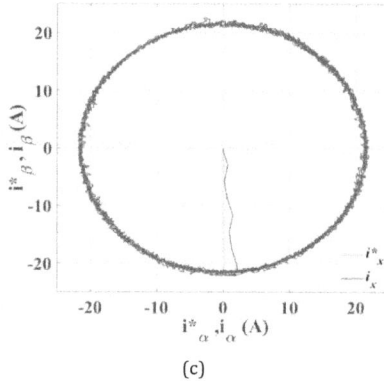

(c)

Figure 3.7: (a) Vector trajectories of $i^*_x$ and $i_x$, for ±10% step changes in grid voltage, at 0.1s and 0.3s respectively, (b) & (c) Vector trajectory of $i^*_x$, and $i_x$, for ±0.5Hz step change in grid frequency at 0.1s and 0.3s respectively.

Even though at the instant of step change a small deviation in trajectory is seen, MPC is able to regain the tracking capability without much delay. Figures 3.7.(b) and (c) show the current vector trajectories for step changes in grid frequency of ±0.5 Hz at 0.1s and 0.3s respectively. The deviations of actual current, $i_x$, at the instant of transition is not much significant and the inverter is still capable of meeting the specified control target with tracking accuracies of 99.6% and 99.92% in magnitude and angle respectively. So, the overall results of the various case studies prove that the proposed MPC based grid-tied inverter exhibits superior dynamic tracking proficiency.

## 3.4.4 Key findings of the sensitivity analysis

(i). The tracking accuracy and THD reduced with increase in sampling frequency.

(ii). The random variation in switching frequency was found to follow a trend of averaging around 1/8th of applied sampling frequency as evident from Table 3.5.

(iii). The current spectrum is spread up to half of sampling frequency even with lower THD for all cases.

(iv). The total harmonic distortion of actual current for any sampling frequency is observed to be within the stipulated IEEE limits.

(v). The inverter power loss escalates with increase in sampling frequency as higher sampling frequencies result in higher switching frequencies which in turn increase the inverter power loss.

(vi). Tracking accuracy improves as sampling frequency increases; at higher sampling frequencies, the predictions being much closer, the reference is better followed.

## 3.5 MPC with bidirectional power flow for IDM

In IDMs, the grid-tied inverters have to function as an interface for power transfer between these AC and DC feeders. Hence, grid-tied inverters in IDMs are mandated to possess bidirectional power flow capability. The second research objective of this thesis is to design and develop MPC based grid-tied inverter with bidirectional power flow feature.

### 3.5.1 The concept of bidirectional power flow in grid-tied inverters

The three phase grid-tied inverter shown in Figure 3.8.(a) can be made to operate as an active front end rectifier when the power transfer is from AC side to DC side and as an inverter when the power transfer is reversed.

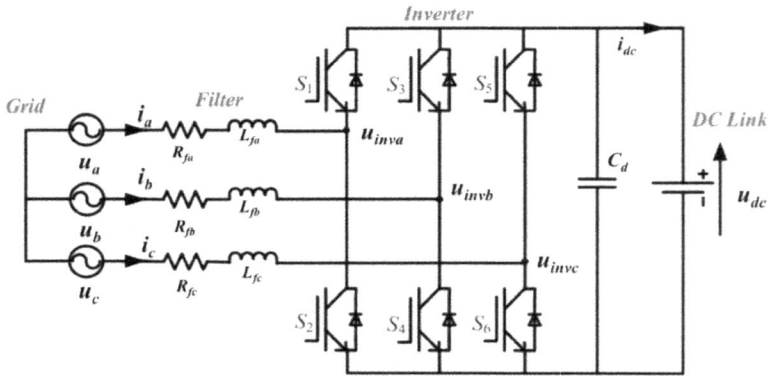

(a)

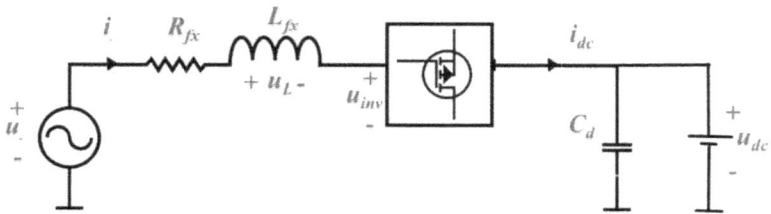

(b)

Figure 3.8: (a) Three phase bidirectional grid-tied inverter system and (b) its Equivalent circuit.

The equivalent circuit of grid-tied inverter [125], [137] connected to grid is shown in Figure 3.8.(b), wherein, $u_x$, and, $i_x$, are the AC side voltage and current, $L_{fx}$, and, $R_{fx}$, are the source/filter inductance and resistance, $u_L$, is the voltage drop across, $L_{fx}$, $u_{inv}$, is the converter pole voltage, $C_d$, is the dc link capacitance and, $u_{dc}$, and, $i_{dc}$, are DC side voltage and currents respectively. Figure 3.9 depicts the phasor diagram [138] for illustrating the concept of bidirectional power flow through a power converter with the system equations of voltage, active power, $P$, and reactive power, $Q$, expressed as,

$$u_x = u_{inv} + u_L = u_{inv} + j\omega L_{fx} i_x \quad .........(3.11)$$

$$P = u_x i_x \cos\theta = \frac{u_s u_{inv}}{\omega L_{fx}} \sin\delta \quad .........(3.12)$$

$$Q = v_s I_{s1} \sin\theta = \frac{v_s^2}{\omega L_s}\left(1 - \frac{v_{c1}}{v_s}\cos\delta\right) \quad .........(3.13)$$

where, $u_x$, $u_{inv}$, $u_L$, and, $i_x$, are the fundamental frequency phasors of grid voltage, inverter voltage, voltage drop across filter and inverter current respectively, $\omega$, is the line frequency in rad/s ,$\theta$, is the phase angle between $u_x$ and $i_x$ and, $\delta$, is the phase angle between $u_x$ and $u_{inv}$ .

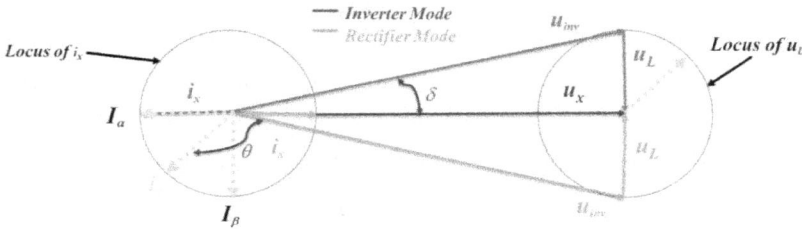

Figure 3.9: General phasor diagram of three phase bidirectional grid-tied inverter.

The inverter control has to ensure (i) $u_{inv}$ lags $u_x$ by an angle $\delta$ in rectifier mode, and (ii) $u_{inv}$ leads $u_x$ by an angle $\delta$ in inverter mode of operation, as demanded by equation (3.11) and (3.12). Correspondingly, the inductor voltage drop, $u_L$, and actual current, $i_x$, change polarity. As per equation (3.11)-(3.12), the inductor voltage drop will be dynamically computed to obtain the inverter reference voltage such that any desired $P$ and $Q$ can be delivered or drawn.

## 3.5.2 Development of direct power MPC scheme

In this research work, a Bidirectional MPC (BMPC) has been formulated and tested for various conditions of IDM. The discrete system model described by equation (3.6) is used here for current prediction while the grid voltage, $u_x(k)$, is estimated through back extrapolation techniques as,

$$u_x(k-1) = u_{inv}(k-1) - \frac{L_{fx}}{T_s}i_x(k) - \left(R_{fx} - \frac{L_{fx}}{T_s}\right)i_x{}^P(k-1) \dots\dots(3.14)$$

The $u_x(k)$ in equation (3.6) can be presumed as $u_x(k-1)$, as grid voltage variations over a sampling period is negligibly small. Hence back propagation estimates the grid voltage for the preceding sample and $u_x(k-1)$ is applied in place of $u_x(k)$ in equation (3.6) to complete the current prediction for $k+1^{th}$ sampling instant. This is then passed on for optimization and further switching state selection.

The features of instantaneous PQ theory [138] is tailored to match the basic MPC to make it a comprehensive controller in power converter outer loops. The BMPC computes the reference current from the reference power based on the Instantaneous PQ theory [138], [140]. The active and reactive powers are expressed on instantaneous stationary reference frame as

$$\begin{bmatrix} p_{\alpha\beta} \\ q_{\alpha\beta} \end{bmatrix} = \begin{vmatrix} u_\alpha i_\alpha + u_\beta i_\beta \\ u_\beta i_\alpha - u_\alpha i_\beta \end{vmatrix} \dots\dots(3.15)$$

where $i_\alpha, i_\beta, u_\alpha, u_\beta, p_{\alpha\beta}$, and, $q_{\alpha\beta}$, are the instantaneous stationary reference frame coordinates of AC current, voltage, active power and reactive power respectively. Thus, the stationary reference frame currents can be expressed as,

$$\begin{bmatrix} i_\alpha \\ i_\beta \end{bmatrix} = \frac{1}{u_\alpha^2 + u_\beta^2} \begin{bmatrix} u_\alpha & -u_\beta \\ u_\beta & u_\alpha \end{bmatrix} \begin{bmatrix} p_{\alpha\beta} \\ q_{\alpha\beta} \end{bmatrix} \dots\dots(3.16)$$

Following equation (3.16), BMPC develops the computation strategy to obtain the reference currents from the reference power.

Implementation of the BMPC computation demands determination of unknown quantities like instantaneous stationary reference frame coordinates of AC voltage, $v_\alpha$ and $v_\beta$, which can be conventionally obtained by direct measurements. Instead, it is proposed here to estimate the voltage parameters utilizing the prediction capability of MPC through equation (3.14). With this reformation MPC becomes a fast, comprehensive, and independent controller for power converters. The single objective weight function represented in equation (3.8) is utilized to deploy the formulated BMPC to achieve power tracking.

## 3.6 Simulation and analysis of the developed BMPC scheme

Having developed the BMPC algorithm, its proof of concept is presented in this section by validating it in realtime applications. Two different applications are considered for testing BMPC:

(i)  Bidirectional power exchange in a microgrid

(ii)  Bidirectional power exchange between battery energy storage systems and grid.

### 3.6.1 Power exchange in a microgrid

BMPC has been employed to accomplish bidirectional power transfer as dictated by a power reference and transfers power from the DC link to microgrid and vice versa. In Figure 3.10, the, $P^*$, and, $Q^*$, are the active and reactive power references for both the rectifier and the inverter modes, and $i^*_x$, is the corresponding reference current generated using equation (3.16). The direction of power flow is decided by the polarity of $P^*$ and $Q^*$. Positive polarity of power references indicates export of power from DC link to microgrid while a negative polarity indicates import of power from microgrid to the DC link.

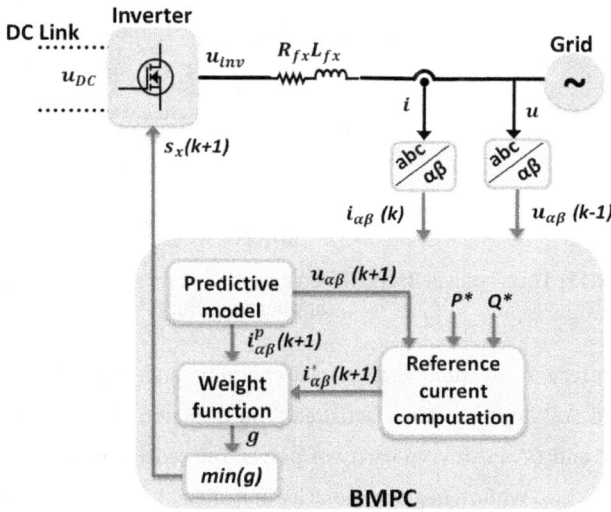

Figure 3.10: Three-phase BMPC based grid-tied inverter for microgrid.

### 3.6.2 BMPC for battery energy storage applications

A three-phase grid-tied inverter with BMPC has been designed for charge/discharge control of a battery in grid interactive energy storage system or an electric vehicle battery. The state of charge of battery, $SoC$, C rating of battery, $C$, duration of charge/discharge operation, $t_d$, and rate of charging, $fast/slow$, are the inputs to reference current generator of the electric vehicle or battery energy storage system as seen in Figure 3.11. These input parameters decide the conditions and limits of the power exchange.

The DC side current is monitored and limited to a value of $I_{dmax}$ to ensure that the battery $C$-rating is not violated in either direction. The relationship among $C$-rating, rate of charging and duration of charge/discharge operation, $t_d$, is described as,

$$I_{dmax} = {C}/{t_d}, I_{dmax} \leq mC \begin{bmatrix} m = 0.1 \ for \ slow \ charging \\ = 0.2 \ for \ fast \ charging \end{bmatrix} \ \dots\dots(3.17)$$

where, $m$ is the multiplication factor suggested by the manufacturer [143].

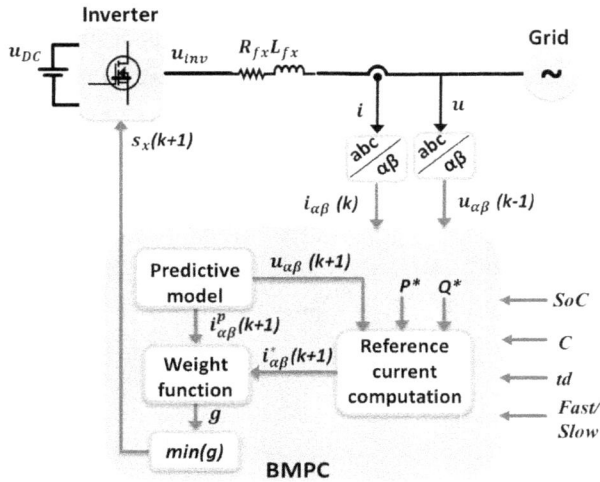

Figure 3.11: Three-phase BMPC based grid-tied inverter for battery energy storage applications.

Thus, the battery protection against overcharging and deep discharging based on the recommended $SoC$ limit has been satisfied. The direction of power flow is decided by the polarity of $P^*$ and $Q^*$. Positive polarity of power references indicates export of power from battery to microgrid while a negative polarity indicates import of power from the microgrid to battery.

### 3.6.3 Simulation of the BMPC inverter

The performance analyses of the BMPC inverter in the applications have been carried out in MATLAB/Simulink platform with the system specifications in Table 3.6. The battery rating and specifications used in this work are as per BYD F3DM battery used in electric vehicle applications [143].

The steady state and the transient performances of the developed BMPC inverter have been assessed using four indices, viz. tracking accuracy, total harmonic distortion of current, $THD_i$, inverter power loss, $P_{loss}$, and the transition time between the modes.

TABLE. 3.6 SYSTEM SPECIFICATIONS

| Parameters | Value | Parameters | Value |
|---|---|---|---|
| DC link voltage | 700 V | Filter inductance | $0.001\Omega$, 10mH |
| Inverter VA rating | 15 kVA | Battery specifications | 700V,180 Ah |
| Grid specifications | 3Φ,400V(l-l),50 Hz | Sampling frequency | 25kHz |

**Case 1:** Microgrid

A microgrid system with various RE sources like solar PV, wind, local loads and battery storage has been considered as the test bench. The developed BPMC algorithm has been applied to control the power flow through the grid-tied inverter of the RE sources in the microgrid. The microgrid system has been realized in MATLAB/Simulink using equivalent impedance and filter impedance along with the grid-tied inverter. The BMPC algorithm has been developed as an embedded function block. The power references have been applied as inputs to the BMPC block. The developed microgrid system has been evaluated as per the test conditions depicted in Table 3.7.

TABLE.3.7 POWER REFERENCES FOR MICROGRID

| Step | Time period (s) | P* (kW) | Q* (kVAR) | Direction of power flow |
|---|---|---|---|---|
| 1 | 0 to 0.2 | 5 | 0 | Export |
| 2 | 0.2 to 0.4 | -5 | 0 | Import |
| 3 | 0.4 to 0.6 | 8 | 2.42 | Export |
| 4 | 0.6 to 0.8 | 5 | 0 | Export |
| 5 | 0.8 to 1 | -5 | 0 | Import |
| 6 | 1 to 1.2 | -10 | -4.48 | Import |

The test cases include (i) steady state in both the modes, (ii) step changes within the modes, and (iii) step change across the modes.

**Case 2:** Battery energy storage applications

In this case, the simulation model considered a battery block from the Simscape Electrical™ library in MATLAB/Simulink. The specifications like *SoC*, C rating of battery, duration of charge/discharge operation of the battery has been modified to represent a commercially available car battery [143]. The battery is connected to the main grid via a bidirectional, three phase grid-tied inverter. The BMPC has been developed through embedded function with inputs like *SoC*, C rating, rate of charging, *fast/slow*, and $t_d$ in addition to the power references. The BMPC performs the charging and discharging through the bidirectional grid-tied inverter while monitoring the $I_{dmax}$ as per equation 3.17, thereby ensuring the safe operating limits of the battery. Table 3.8 presents the set of active and reactive power references applied in case 2. The same performance parameters as evaluated in case 1 are considered even for this case.

TABLE.3.8 POWER REFERENCES AND OTHER INPUTS FOR CASE 2

| Step | Time period (s) | P* (kW) | Q* (kVAR) | Duration for charge/discharge operation, td, (s) | Charging option (fast/slow) | Direction of power flow |
|------|------|------|------|------|------|------|
| 1 | 0 to10 | 5 | 0 | 10 | Slow | Export |
| 2 | 10 to 20 | -5 | 0 | 10 | Slow | Import |
| 3 | 20 to 30 | 8 | 2.42 | 10 | Fast | Export |
| 4 | 30 to 40 | 5 | 0 | 10 | Slow | Export |
| 5 | 40 to 50 | -5 | 0 | 10 | Slow | Import |
| 6 | 50 to 60 | -10 | -4.48 | 10 | Fast | Import |

## 3.6.4 Steady state performance analysis of BMPC inverter

**Case 1:** Microgrid

The simulation results of case 1 presented in Table 3.9 along with Figures 3.12 (a) and (b) indicate a commendable tracking accuracy of around 99% in active power and 97% in reactive power for all conditions in both the modes.

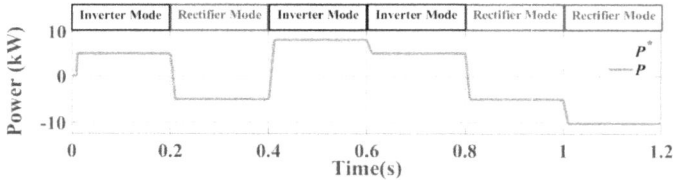

(a)

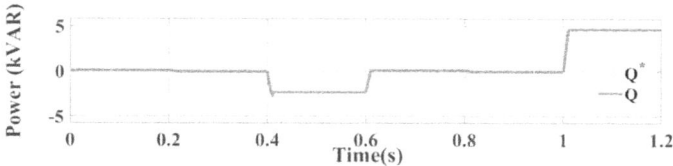

(b)

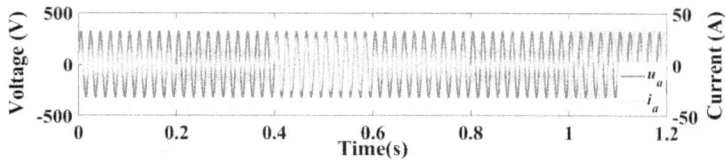

(c)

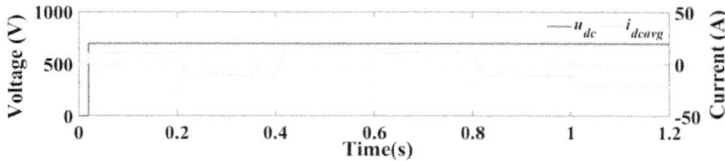

(d)

Figure 3.12: Time Response of BMPC in microgrid with step changes at every 0.2s in: (a) Active power reference, $P^*$, (b) reactive power reference, $Q^*$, (c) grid voltage, $u_a$, and actual current of phase a, $i_a$, (d) DC link voltage, $u_{dc}$, and average current, $i_{dcavg}$.

TABLE.3.9 PERFORMANCE INDICES OF BAMPC FOR MICROGRID

| Case 1 | Tracking accuracy in % | | $THD_i$ (%) | Converter power loss $P_{loss}$ (%) |
|---|---|---|---|---|
| | P | Q | | |
| | 99.48 | 97.18 | 2.94 | 0.38 |
| | 99.46 | 97.18 | 2.92 | 0.3 |
| Microgrid | 99.42 | 98.1 | 2.86 | .98 |
| | 99.48 | 97.5 | 2.96 | 1.16 |
| | 99.46 | 97.3 | 2.94 | .46 |
| | 99.42 | 98.6 | 2.56 | 1.43 |

Figure 3.13 depicts the voltage and current waveforms in "inverter mode" and "rectifier mode" of operation of the grid-tied inverter in the present case. Figure 3.13.(a) shows the upf operation in inverter mode described in step 1 of Table 3.7. Figures 3.13 (b), (c) and (d) present the rectifier mode of step 2, rectifier mode of step 6 and inverter mode of step 3 respectively. The amplitude of AC side currents in Figure 3.13.(c) has been found changing in response to change in $P^*$, and the phase angle of current is found varying in accordance with $Q^*$.

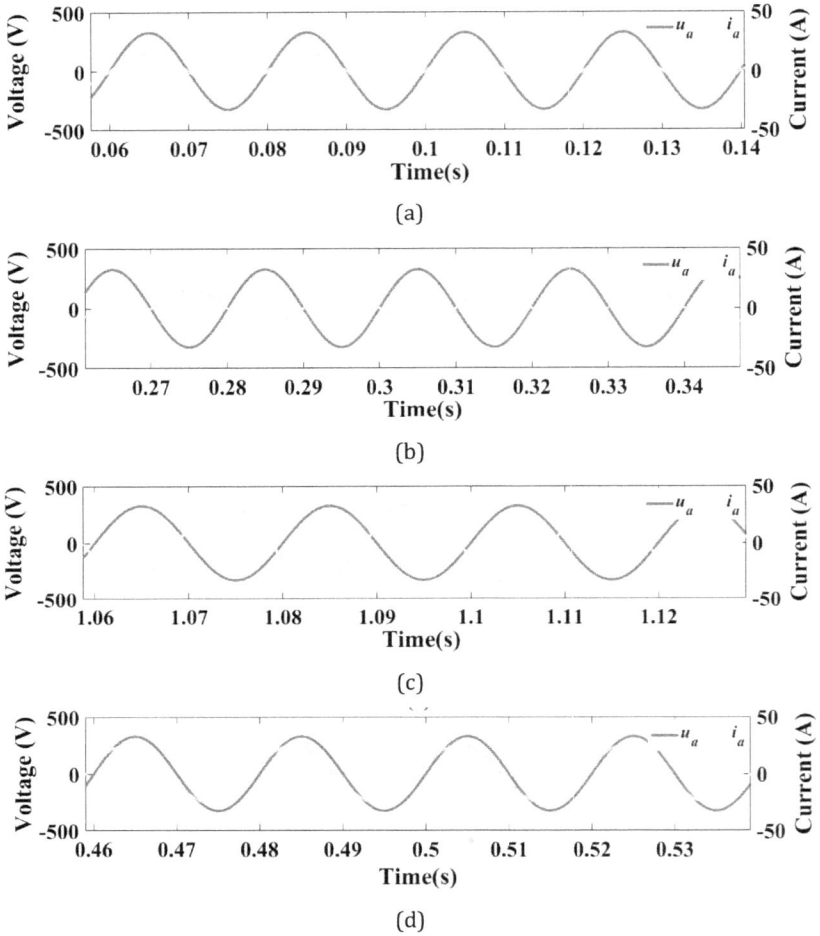

Figure 3.13: Time response for grid voltage, $u_a$, and actual current, $i_a$, of phase A for BMPC applied to the microgrid: (a) and (d) Inverter mode of operation; (b) and (c) Rectifier mode of operation.

During the rectifier mode of operation at upf, the inverter current is in phase opposition with respect to the grid voltage, as seen from Figure 3.13. (b). In Figure 3.13.(c) and (d), the current is seen to lead or lag the voltage in accordance with Q*.

Figure 3.14 gives various system variables of the grid-tied inverter during the mode transition from inverter to rectifier mode and vice versa.

84

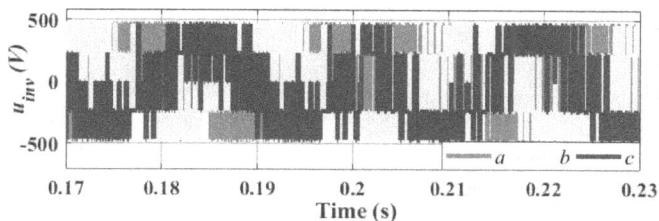

(a)

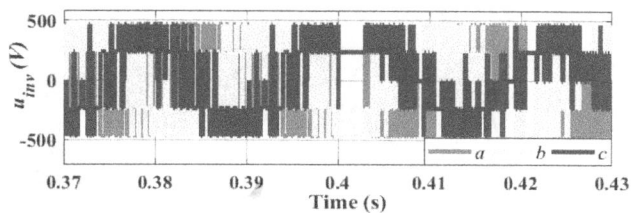

(b)

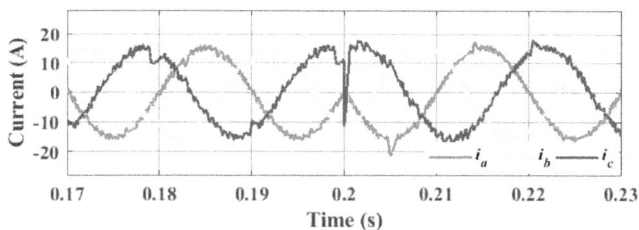

(c)

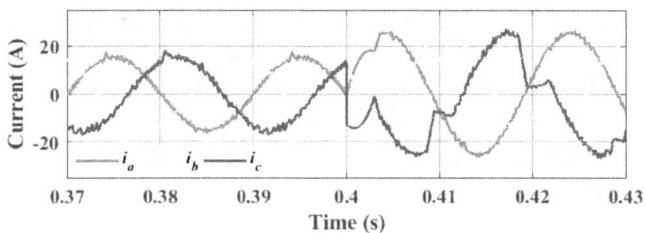

(d)

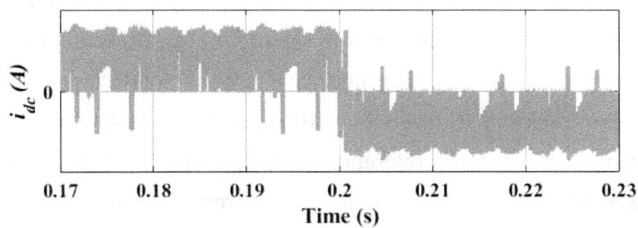

(e)

85

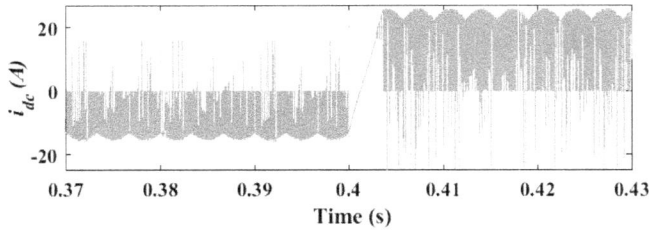

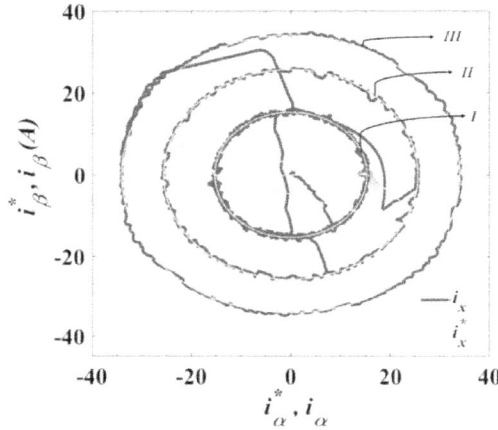

Figure 3.14: Time Response of BMPC for microgrid during the mode transition: (a) $u_{inv}$ for step 1 to step 2; (b) $u_{inv}$ for step 2 to step 3 (c) load current for step 1 to step 2; (d) load current for step 2 to step 3; (e) DC side current for step 1 to step 2; (f) DC side current for step 2 to step 3; (g) trajectories of $i_x$ and $i^*_x$.

During transition between inverter and converter modes, the following observations are made:

(i)     The DC current polarity changes.

(ii)    Phase of the AC current reverses.

(iii)   The phase sequence of converter pole voltage alternates between abc and acb.

(iv)    The highly accurate reference tracking capability of BMPC is observed from the current vector trajectories I, II and III illustrated in Figure 3.14.(g). Trajectory I correspond to a $P^*$ of 5 kW (steps 1, 2, 4 and 5) and trajectory II is for a $P^*$ of 8 kW and 2.42 kVAR (step 3) respectively. Trajectory III had a $P^*$ of 10 kW and a $Q^*$ of 4.48 kVAR (step 6). The actual current, $i_x$, tracked the reference current, $i^*_x$, perfectly with the trajectories superimposed on each other for step change in power references.

86

(v)     The $THD_i$ is observes to be below 3% (Table 3.9) meeting the IEEE standards. The converter power loss is found to be only 1.43 % (Table 3.9) at the specified sampling frequency of 25 kHz.

**Case 2:** Battery Energy Storage Applications

In this case, the BMPC has similar steady state characteristics as in the previous test case with almost the same reference tracking accuracy, low $THD_i$, power loss within 1.53% as presented in Table 3.10.

TABLE.3.10 PERFORMANCE INDICES OF BMPC IN BATTERY ENERGY STORAGE APPLICATIONS

| Case 2 | Tracking accuracy in % | | Current harmonic profile ($THD_i$) | Converter power loss ($P_{loss}$) in % |
|---|---|---|---|---|
| | Active power | Reactive power | | |
| Battery Energy Storage | 99.49 | 97.10 | 2.93 | 0.4 |
| | 99.4 | 97.10 | 2.92 | 0.34 |
| | 99.3 | 98.08 | 2.80 | 1 |
| | 99.49 | 97.3 | 2.39 | 1.2 |
| | 99.38 | 97.3 | 2.92 | .48 |
| | 99.24 | 98.6 | 2.48 | 1.53 |

Figures 3.15. (a) and (b) show that the actual power is in good compliance with the $P^*$. Figure 3.15.(d) shows the rise and fall of battery $SoC$ in accordance with the modes of operation of the inverter. The $SoC$ of the battery rises when the power flows into the battery and falls when the power is delivered by the battery.

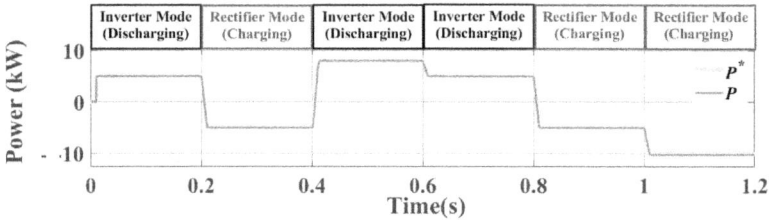

(a)

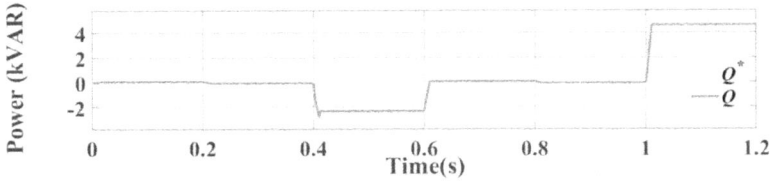

(b)

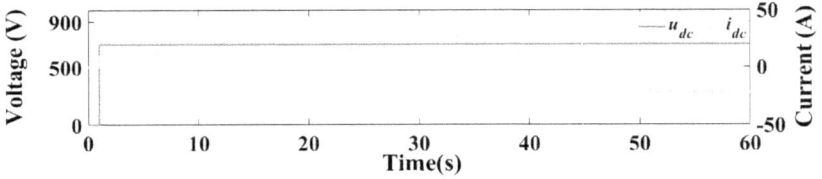

(c)

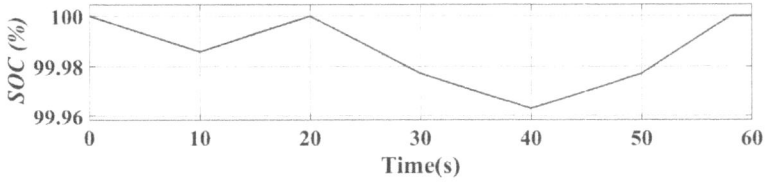

(d)

Figure 3.15.: Time Response of BMPC for Case 2 with step change at every 10min as shown in Table 3.8: (a) $P^*$; (b) $Q^*$; (c) DC link voltage, $u_{dc}$, and average DC current, $i_{dcavg}$. (d) $SoC$.

The plots in Figures 3.17 show the AC side current and voltage per phase during the mode transitions. Similar to the previous test case, the current phase reversal when moving from inverter to rectifier mode is also seen for this case as well. Figures 3.17 (a), (b), (c) and (d) present various system quantities during steps 1, 2, 3 and 6 of Table 3.8.

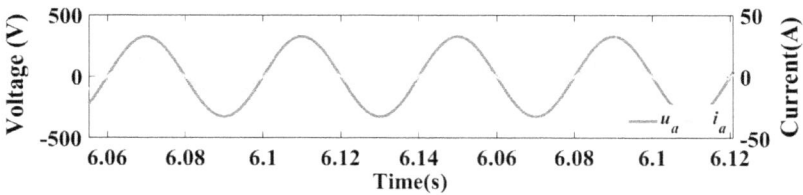

(a)

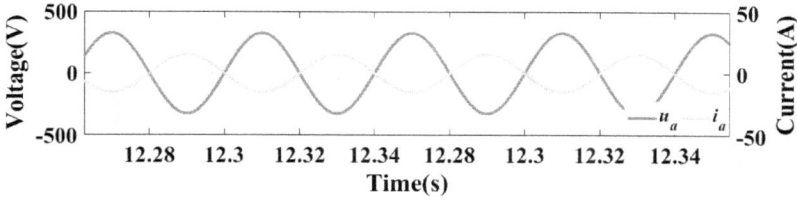

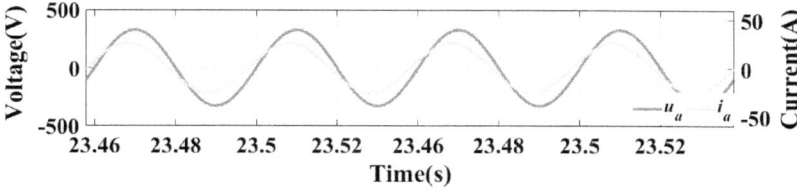

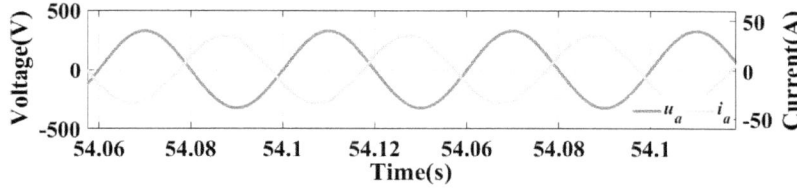

Figure 3.16: Time response of grid voltage, $u_a$ and inverter current, $i_a$ of phase A with BMPC inverter for case 2: (a) and (c) Inverter mode of operation; (b) and (d) Rectifier mode of operation.

Figure 3.17 depicts the system variables at the instants of transition from inverter mode to rectifier mode and vice versa. Similar trends with respect to the phase sequences of currents, inverter voltages and the current trajectories are observer before and after transition.

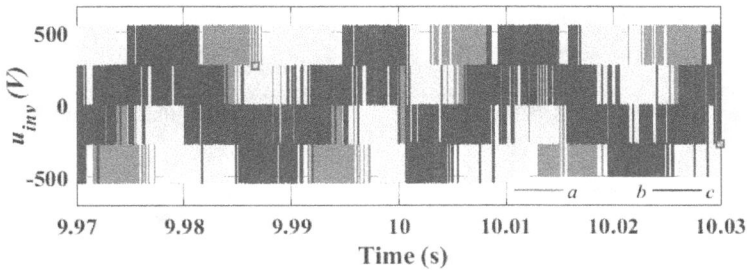

(a)

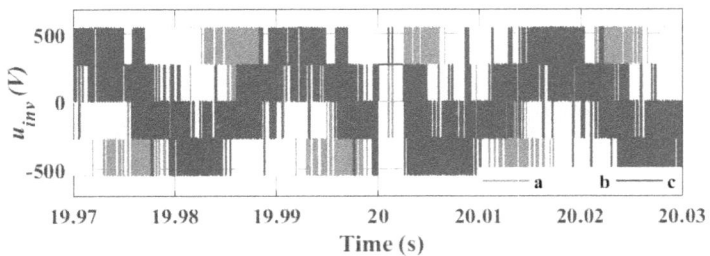

(b)

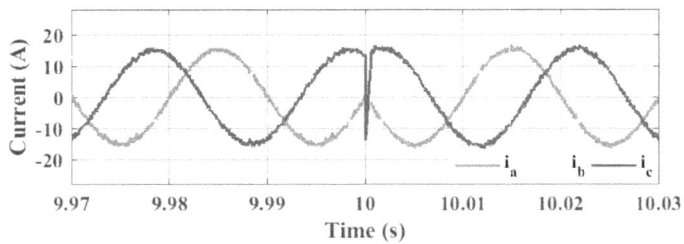

(c)

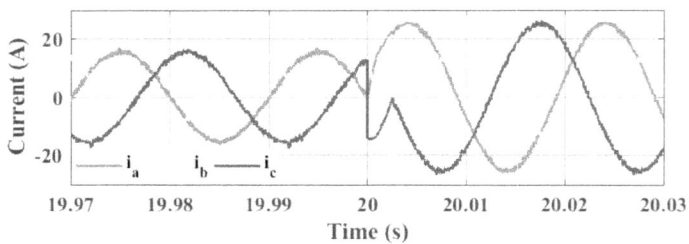

(d)

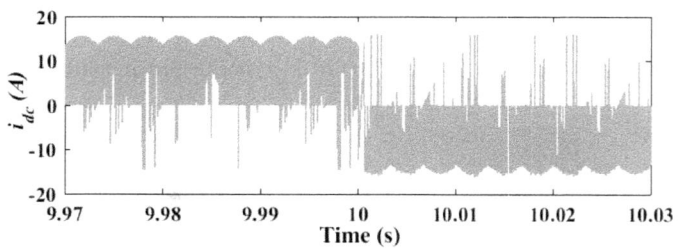

(e)

90

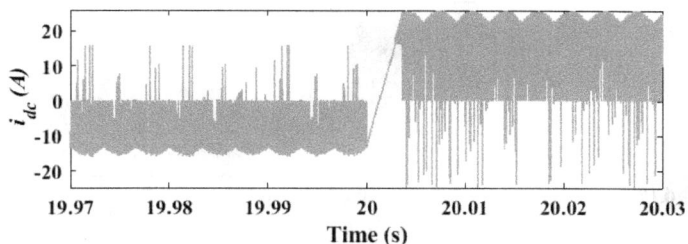

(f)

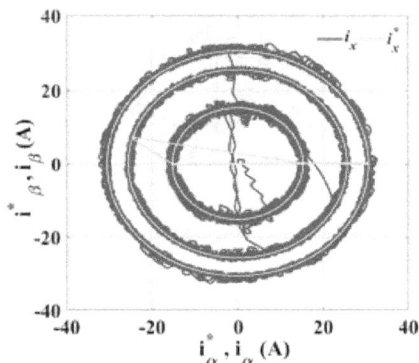

(g)

Figure 3.17: Time Response of BMPC for Case 2 at mode transitions: (a) & (d) inverter pole voltages, $u_{inv}$, (b) & (e) actual current, $i_x$, (c) & (f) DC side current, $i_{dc}$ and (g) Current Trajectory of $i_x$ and $i^*_x$.

## 3.6.5 Transient analysis of BMPC inverter

Performance of BMPC based inverter under transient events is validated by observing the transition times under two conditions:

(i) mode change from inverter to rectifier, and, (ii) step change in power references.

Figure 3.18, Figure 3.29 and Table 3.11 presents the transition edge for both microgrid and battery energy storage applications.

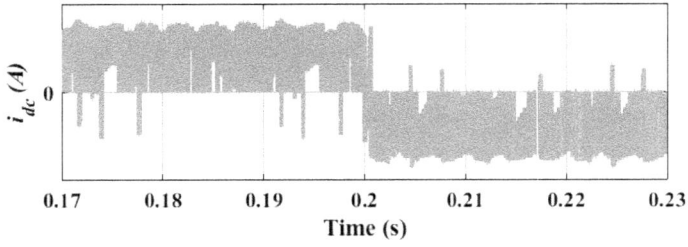

(a)

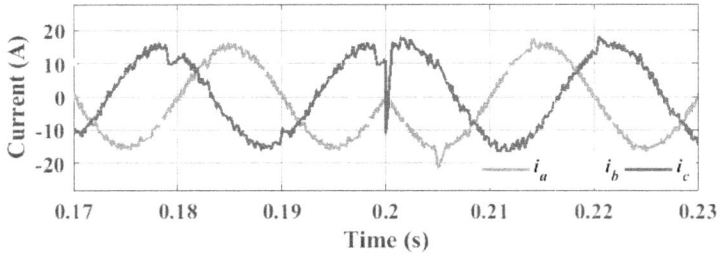

(b)

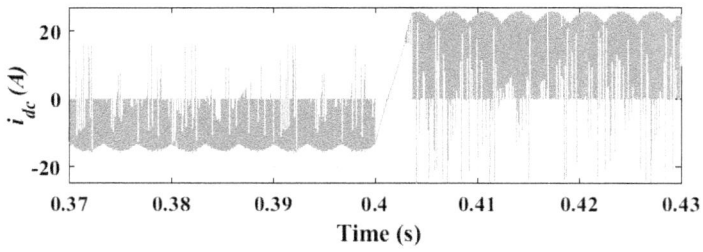

(c)

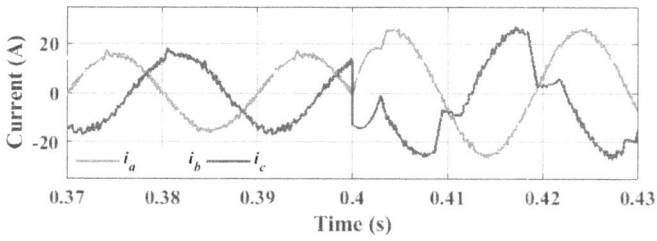

(d)

92

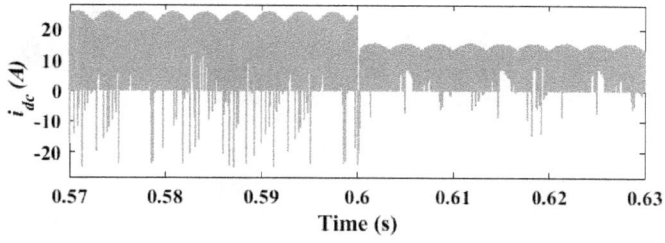

(e)

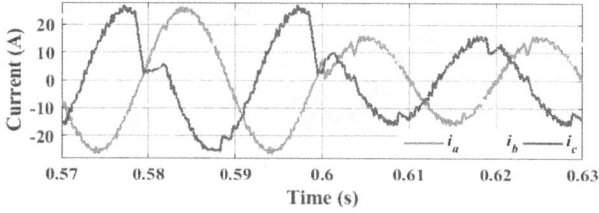

(f)

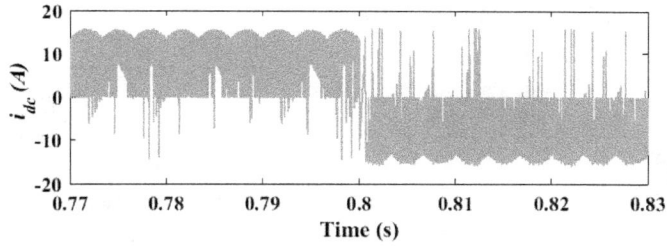

(g)

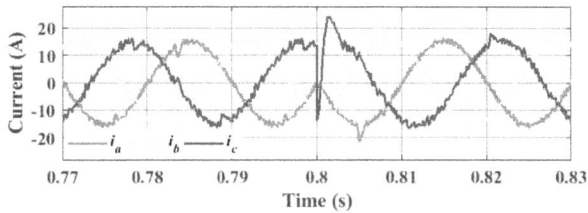

(h)

93

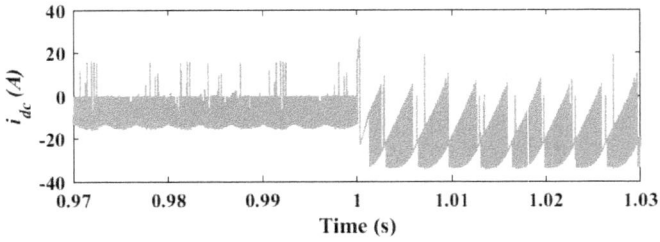

(i)

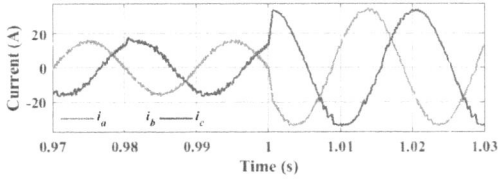

(j)

Figure 3.18: Current transitions of BMPC inverter in microgrid: (a), (c), (e), (g) and (i) DC side current, $i_{dc}$; (b), (d), (f), (h) and (j) AC side currents, $i_x$.

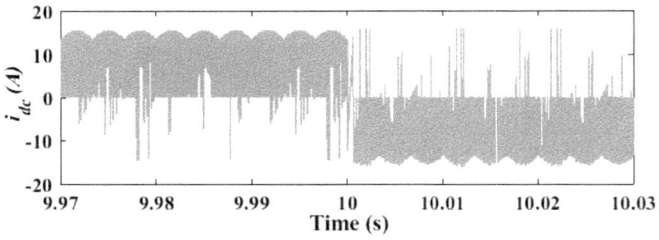

(a)

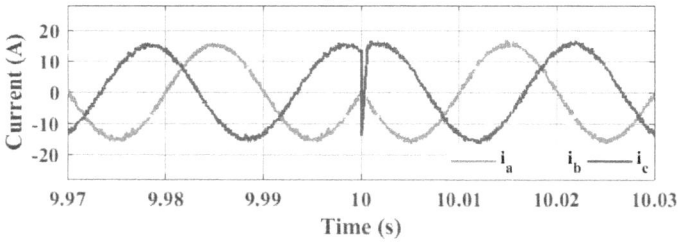

(b)

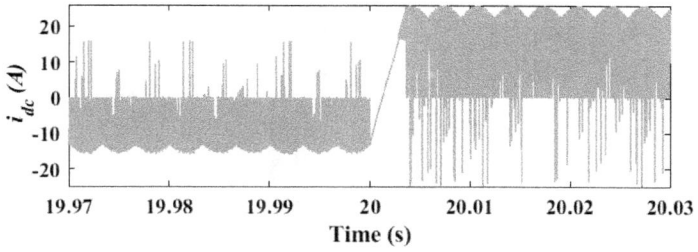

(c)

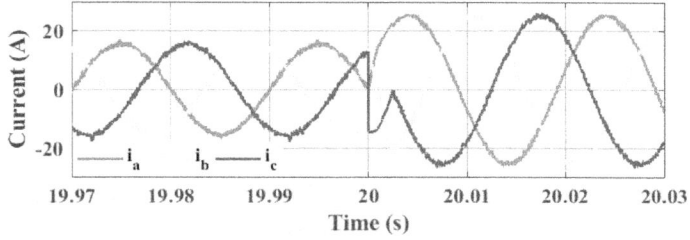

(d)

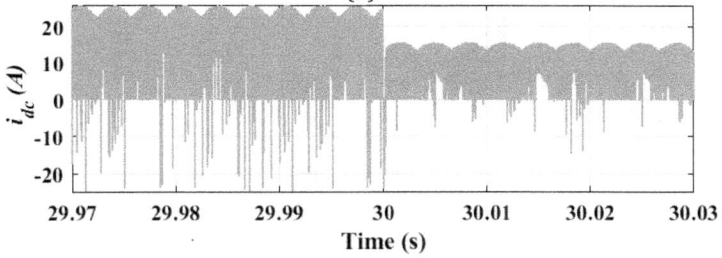

(e)

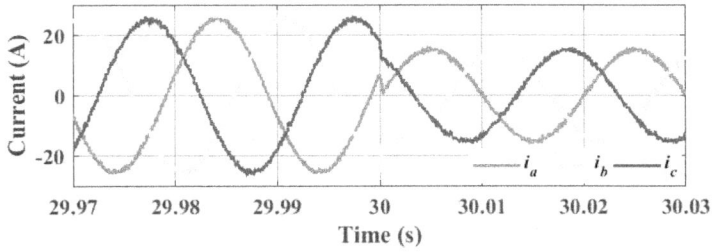

(f)

95

(g)

(h)

(i)

(j)

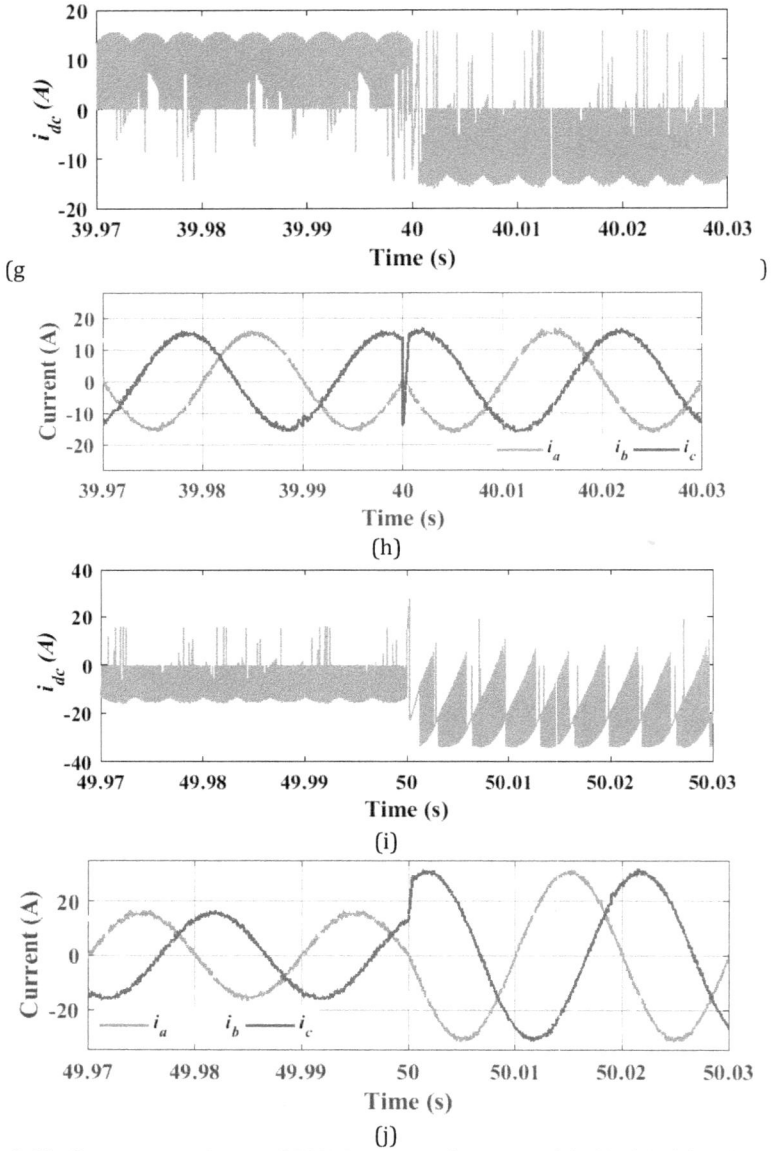

Figure 3.19: Current transitions of BMPC inverter for Case 2:(a), (c), (e), (g) & (i) DC side currents, $i_{dc}$; (b), (d), (f), (h) & (j) AC side current, $i_x$.

From Figures (3.18) to (3.19) it is observed that BMPC takes 0.4 ms for mode transition alone and 0.5 ms for step changes in $P^*$ & $Q^*$ with or without mode change.

TABLE.3.11. TRANSITION TIMES of BMPC

| Transition condition | Power references | Transition time (ms) |
|---|---|---|
| Mode change from Inverter to Rectifier | $P^*$=5 kW, $Q^*$= 0 kVAr | 0.4 |
| Mode change from Rectifier to Inverter | $P^*$= 5 kW & $Q^*$= 0 kVAr | 0.4 |
| Step change in both $P^*$ & $Q^*$ within inverter mode | $P^*$ from 5 to 8 kW & $Q^*$ from 0 to 2.42 kVAR | 0.5 |
| Mode change from Inverter to Rectifier and step change in both $P^*$ & $Q^*$ | $P^*$ from 8 kW to 5 kW & $Q^*$ from 2.42 kVAR to 0 | 0.5 |

### 3.6.6 A summary on the bidirectional MPC

The salient aspects from the analysis of the BMPC can be summarized as:

(i)     The mode transition time taken to move between inverter and rectifier modes increases by approximately 25% when power references are also changed simultaneously.

(ii)    The AC current amplitude is proportional to active power demand and it increases when active power demand increases.

(iii)   The AC current phase angle with respect to grid voltage varies as per reactive power demand. The currents are in phase with voltage for upf mode.

(iv)    The DC current polarity reverses when moving from inverter and converter mode but DC side voltage polarity remains the same.

(v)     The control is capable of tracking step changes in active and reactive power references without much loss in accuracy but with slightly larger transition times.

## 3.7 Summary

This chapter started with a detailed sensitivity analysis of sampling frequency on the performance of MPC based grid-tied inverter. Based on this analysis, the switching frequency of inverter MPC was found to follow a trend of averaging around 1/8th of applied sampling frequency.

Next, the design and development of Bidirectional MPC has been presented and tested for two applications viz. grid-tied inverters in microgrid and charge/discharge control of battery energy storage applications. Bidirectional power control, elimination of grid measurement and improved dynamic responses in the range of 0.4 ms to 1.53 ms are the major advantages of the developed bidirectional MPC. This study completes the investigation on the second research objective of this thesis.

# Chapter 4

# Development of Dynamically Adaptable Model Predictive Control for Inverter Dominated Microgrids

## 4.1 Introduction

This chapter presents the design and development of dynamically adaptable MPC to work in IDMs. The chapter start with the feasibility study on the influence of the multi objective optimization on the performance of MPC. It further details the advanced control features and the importance of self-adaptability in inverter controls. The test results of simulation as well as implementation in hardware for a dynamically adaptable MPC in a typical IDM environment are presented as the last section.

## 4.2. Grid-forming control with MPC inverter

Often grid tied inverters working in microgrids shut their operation when islanded, resulting a loss of grid-coupled inertia which disrupt the supply demand balance. But, the paradigm shift in the power system from synchronous generation to inverter-dominated generation, demands every inverter-coupled power unit to contribute to the inertia every time. Grid Forming scheme for inverters adds the capability of direct control of inverter terminal voltage to form the grid independently inspite of the absence of grid voltage [2], [15], [28]. This is identified as an advanced inverter function to work in the future power grids to ensure enhanced reliability. In this work a Grid-Forming Control (GFC) scheme is formulated to impart such a grid-forming capability to the (bidirectional) MPC grid-tied inverter.

Integrating such an independent grid-forming control along with the default grid following control to a MPC based grid-tied inverter is the research work presented below. Further, it is intended to test for its mode transfer ability to move from grid forming to grid tied and vice versa in an IDM system.

## 4.2.1 Modelling and design of grid-forming control of inverters with MPC

The Grid Forming Controller (GFC) developed for MPC based inverter is shown in Figure 4.1.

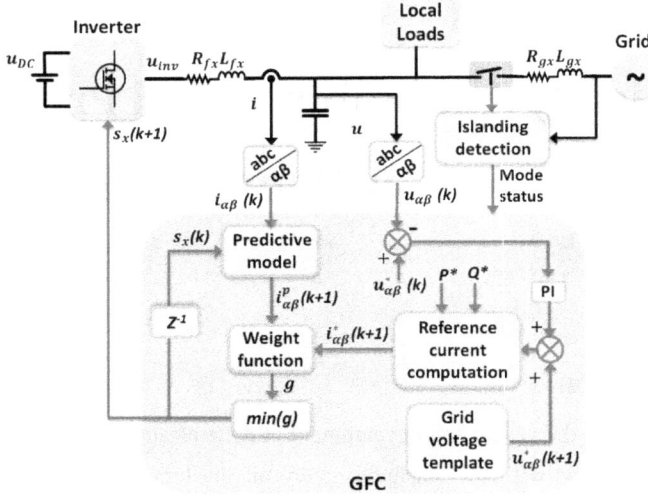

Figure 4.1: Grid-forming control for MPC inverter

It includes an islanding detection unit and the MPC based GFC, the former senses the status of the main grid, while the latter starts its control action after that. When islanded, the mode status signal is enabled and the GFC comes into action. In this research work, a reference signal with the information of rated voltage and rated frequency is pre-defined in the GFC. Simultaneously, the active power reference, $P^*$ is assigned as required by the local demand and can adapt to any dynamically varying values whenever the power balance in the microgrid is to be ensured. The predictive model for GFC and the multi objective weight function to reduce switching frequency has also been formulated as briefed in chapter 3 and is repeated here with the variables of this new inverter system.

The inverter output voltage, $u_{inv}$, based on the load model parameters in GFC system is expressed as,

$$u_{inv} = L_{eqx}\frac{di_x}{dt} + R_{eqx}i_x + u_x \quad ; \{x = a, b, c\} \dots\dots\dots(4.1)$$

where $i_x$ is the per phase inverter current, $u_x$ is per phase grid voltage, and, $L_{eqx} = L_{fx} + L_{gx}$ and $R_{eqx} = R_{fx} + R_{gx}$ are the per phase equivalent inductance and resistance respectively. The suffixes $f$ and $g$ indicate filter and grid impedance components respectively; $x$ indicates the phase. In GFC, the per phase grid voltage template is embedded within the controller as,

100

$u_{\alpha\beta}(k+1)$, as shown in Figure 4.1. The reference, $u_{\alpha\beta}(k+1)$, is in the stationary reference frame with rated amplitude and frequency of grid voltage.

Following the same procedure as in equation (3.1) to (3.6) of chapter 3, the predicted current for $k+1^{th}$ instant can be obtained as,

$$i_x{}^P(k+1) = \left(1 - \frac{R_{eqx}T_s}{L_{eqx}}\right)i_x(k) + \frac{T_s}{L_{eqx}}\left(u_{inv}(k) - u_{\alpha\beta}(k+1)\right) \dots\dots (4.2)$$

As described in chapter 3, the sampling frequency, $F_s$, is purely designer's choice, but from equation (4.2) it is observed to impact the predicted current magnitudes which in turn can affect the tracking accuracy. Selection of sampling frequency demands an inevitable design tradeoff between the reference tracking accuracy and the inverter losses. But, in this work the multi-objective weight function of MPC is utilized to target reduction in switching frequency irrespective of the sampling frequency.

The multi-objective weight function, $g$, with a secondary target added for reduction in switching frequency has been considered and expressed as,

$$g = min \left(\left|i_\alpha^*(k+1) - i_\alpha^P(k+1)\right| + \left|i_\beta^*(k+1) - i_\beta^P(k+1)\right|\right.$$
$$\left. + \lambda\sum\left|s_x^P(k+1) - s_x(k+1)\right|\right) \quad \dots\dots\dots (4.3)$$

where $i_\alpha^*$, $i_\beta^*$, $i_\alpha^P$ and $i_\beta^P$ are the stationary reference frame coordinates of the reference current, $i_x^*$, and the predicted current, $i_x^P$, respectively, $s_x^P$ $(k+1)$ are the status of the top switches of the inverter legs predicted for $k+1^{th}$ instant, $s_x(k)$ are the same for $k^{th}$ instant and $\lambda$ is the secondary function weightage [144].

The value of $\lambda$ decides the relative degree of attainment of the secondary objective with respect to the primary objective. The secondary function of equation (4.3) is formulated to reduce inverter switching frequency for any fixed sampling frequency. Here, the MPC optimization is intended to identify an inverter state which satisfies both the objectives as bounded by $\lambda$. The designer can choose different values of $\lambda$ depending on the allowed tolerance between the tracking accuracy and the intended reduction in the switching frequency. A low $\lambda$ will give more weight on the primary objective than the secondary and can result in a strict tracking; but the switching frequency reduction may be left unattained. Conversely, a high $\lambda$ may result in reduced switching frequency, but compromising on the accuracy of reference tracking and may cause a complete loss of tracking under some operating conditions.

## 4.2.2. Simulation analysis of the grid-forming control for MPC grid-tied inverter

The GFC designed and developed has been validated through simulation studies in MATLAB/Simulink platform. The IDM test bench details and inverter specifications considered for the simulation analysis are described below.

**Test bench description**

A 5-bus IDM emulator available in the Renewable Energy laboratory of Amrita School of Engineering, Coimbatore, India, has been used as the test bench for validation of the GFC. The single line diagram of the IDM emulator has been given in Figure 4.2.

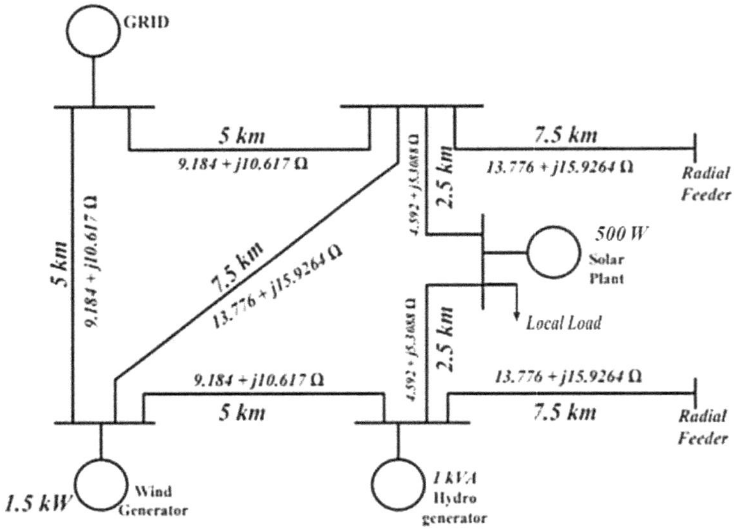

Figure 4.2: Single line diagram of the IDM emulator

The IDM emulator consists of a 1.5 kW wind turbine generator, 1 kVA hydro electric generator, 500 W solar PV array, six distribution lines, and two radial load feeders [18]. This IDM is connected to grid for export or import of power. The MPC based inverter of the solar PV plant with the specifications in Table 4.1 are chosen for validation of GFC.

The equivalent grid impedance as viewed from the inverter output, $R_{eqx}$ and $L_{eqx}$, varies with changes in generation mix (*Genmix*) as presented in Table 4.2. The entry 1 in Table 4.2 indicates presence of the respective power source while 0 indicates its absence [18].

TABLE 4.1 MPC INVERTER SPECIFICATIONS

| Parameters | Value |
|---|---|
| DC link voltage | 700 V |
| Inverter VA rating | 15 kVA |
| Grid specifications | 3φ, 415V(L-L)+ ±10%, 50 ± 0.5 Hz |
| Filter impedance ($R_{fx}, L_{fx}$) | 0.001 Ω, 10 mH |
| Grid impedance ($R_{gx}, L_{gx}$) | 3.093 Ω, 4.4 mH |

TABLE 4.2 EQUIVALENT IMPEDANCE VIEWED FROM INVERTER PORT OF IDM

| Generation mix | Grid | Wind generator | Hydro generator | Local load | Impedance at solar plant output (ω) |
|---|---|---|---|---|---|
| Genmix 1 | 1 | 1 | 1 | 1 | 3.094+j4.531 |
| Genmix 2 | 1 | 0 | 1 | 1 | 3.144+j4.59 |
| Genmix 3 | 1 | 1 | 0 | 1 | 5.334+j7.192 |
| Genmix 4 | 0 | 1 | 1 | 1 | 3.262+j4.731 |
| Genmix 5 | 0 | 0 | 1 | 1 | 3.61+j5.146 |
| Genmix 6 | 1 | 0 | 0 | 0 | 7.486+j9.742 |

*Genmix 1* represents a condition when all the sources in the IDM are present along with the main grid. The MPC inverter will be in grid-tied mode of operation for *Genmix 1* and will synchronously feed power to the main grid. When the main grid is absent, its status becomes '0' in Table 4.2, then the IDM moves to *Genmix 4*. In this mode, the microgrid is islanded and the mode status signal of the GFC enables the MPC inverter to switch to the grid-forming control. The MPC inverter now forms the local grid and delivers the local demand.

The GFC is tested for various *Genmix* conditions of Table 4.2 in MATLAB/Simulink. The following performance indices of the inverter have been obtained:

(i)     average switching frequency, $F_{asw}$ in kHz

(ii)    inverter current harmonics, $THD_i$ in %

(iii)   lowest order harmonics, *LOH*

(iv)    rms current error, $\varepsilon_i$, in %

(v)     rms voltage error, $\varepsilon_u$, in %

(vi)    voltage harmonics, $THD_u$ in %.

Definitions of all the indices are the same as that of chapter 3, but *LOH* is the new index which is added in this chapter. *LOH* is defined as the order of the harmonic, which is closest to the

103

fundamental, thus a higher value of *LOH* represents better quality of power. i.e. high *LOH* means the harmonics are shifted away from the fundamental. This index is added because MPC works with varying switching frequency and varying harmonics orders, so the one which is lowest decides the size of the filter.

Under all these cases, the GFC established the local grid with rated grid voltage and frequency. The *THD$_i$* of the inverter current also adhered to the grid codes. Some salient test results have been discussed below.

The GFC has been tested when IDM is operating in *Genmix 4* and delivering a power of 6 kW at 0.8pf with $\lambda$=0.2 and a sampling frequency, *F$_s$*, of 25 kHz. As seen in Figure 4.3.(a), the developed GFC is observed to establish a local grid as intended.

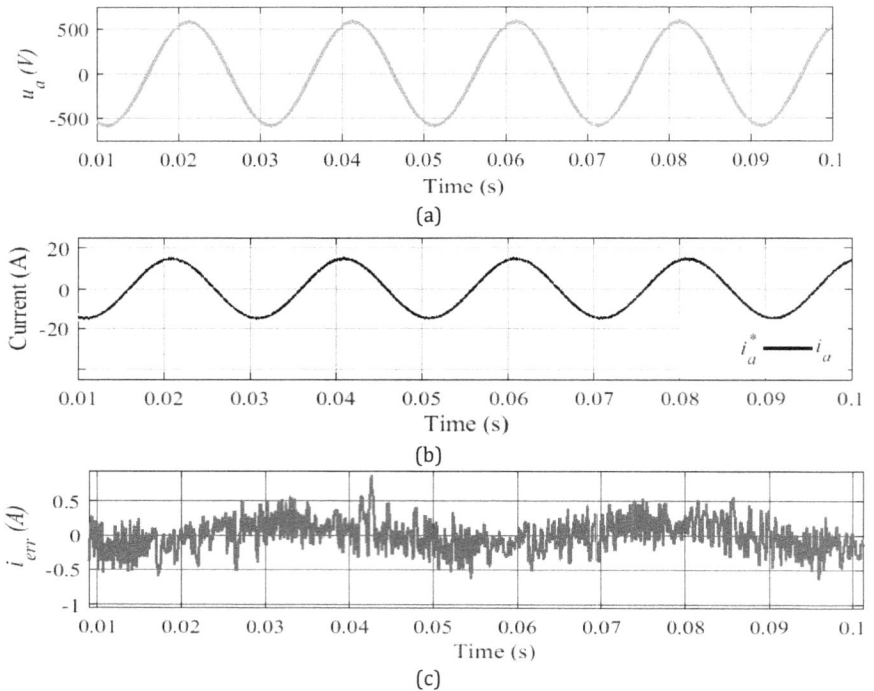

Figure 4.3: Time response of grid-forming controller in simulation with *Genmix 4*: (a). Local grid voltage, *u$_x$*; (b). reference current, *i*$^*_a$, and actual current, *i$_a$*; (c) instantaneous current error, *i$_{err}$*.

The actual current, *i$_a$*, is observed to closely track the reference current, *i*$^*_a$, as depicted in Figure 4.3.(b). The instantaneous current error, $\varepsilon_i$ is seen to be much less as observed in Figure 4.3.(c). The performance indices for this test case have been summarized in Table 4.3.

TABLE 4.3. PERFORMANCE INDICES OF MPC GRID-FORMING INVERTER

| Quantity | Value |
|---|---|
| $i^*(A)$ | 10.06 |
| $i(A)$ | 10.072 |
| $F_{asw}$ (kHz) | 1.48 |
| $THD_i$ (%) | 1.74 |
| LOH | 64 |
| $\varepsilon_i$ (%) | 1 |
| $u^*$ (V) | 239.6 |
| $u$ (V) | 239.6 |
| $\varepsilon_v$(%) | 0 |
| $THDu$ (%) | 1.04 |

## 4.2.3. Hardware validation of the grid-forming control of the MPC inverter

The developed GFC is also validated in real time using the OPAL-RT, OP4200 Hardware-in-Loop (HIL) emulator, whose block diagram is presented in Figure 4.4. The GFC is coded into OPAL-RT, which takes in the grid status and actual current as inputs and generates the gating signals for the inverter. The GFC is tested for various *Genmix* conditions of Table 4.2 in hardware as well. The same set of performance indices of the inverter described for simulation studies have been obtained in hardware validations as well.

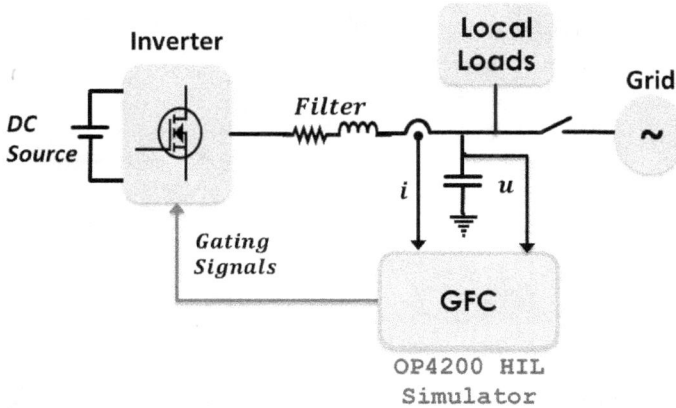

Figure 4.4: Hardware test bench for validation of GFC

The GFC has been establishing the local grid with rated grid voltage and frequency in hardware validations. The $THD_i$ of the inverter current adhered to the grid codes. All the testing of the GFC has been performed in hardware with a switching frequency, $F_s$, of 10 kHz, due to the sampling restrictions of OP4200 HIL emulator. Yet, the GFC has been observed to exhibit a competent performance across all the test cases. A few test results have been chosen from

105

among these and discussed below. In all the figures, voltage is shown in 1:20 scale and currents are shown in 1:2 scale. The instantaneous current errors, gating signals and phase angles are in 1:1 scale.

The first test condition considered is when the IDM is in *Genmix 4* and delivering a power of 6 kW at 0.8pf with $\lambda=0.2$ and the results are presented in Figures 4.5.(a) to (g). The established grid voltage, and the currents are exhibiting high power tracking accuracy. The reduction in inverter switching due to inclusion of a secondary target with $\lambda$ of 0.2 can be observed from the gating signals of Figure 4.5.(g).

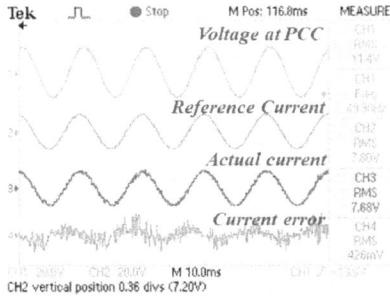

(a).1: Voltage at PCC, 2: reference current, 3: actual current and 4: instantaneous current error, $i_{err}$.

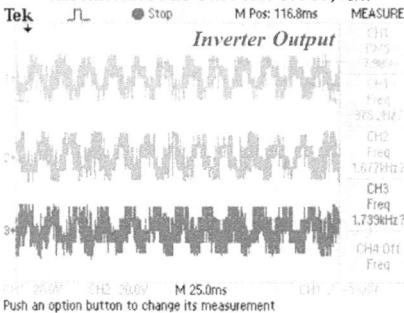

(b). Inverter output voltages- 1: Leg a, 2: Leg b and 3: Leg c.

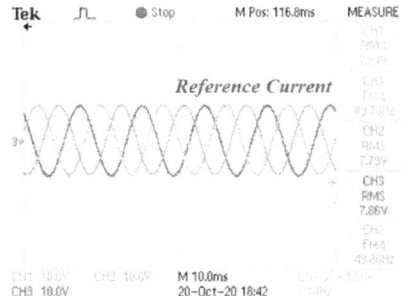

(d). Reference currents for phase a, phase b and phase c.

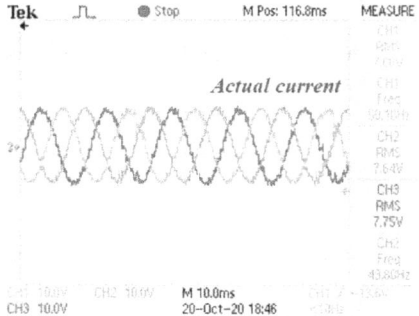

(e). Actual currents for phase a, phase b and phase c.

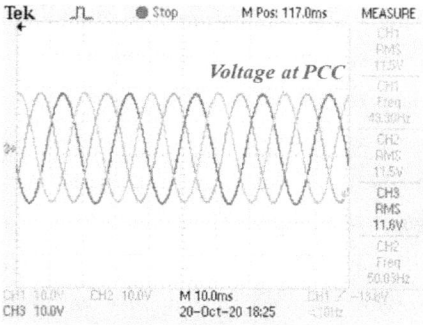

(c). Local grid voltages of phase a, phase b and phase c.

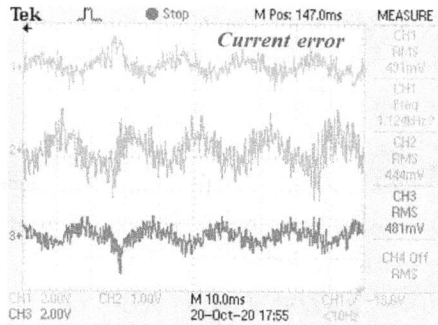

(f). Instantaneous current errors- 1: phase a, 2: phase b and 3: phase c.

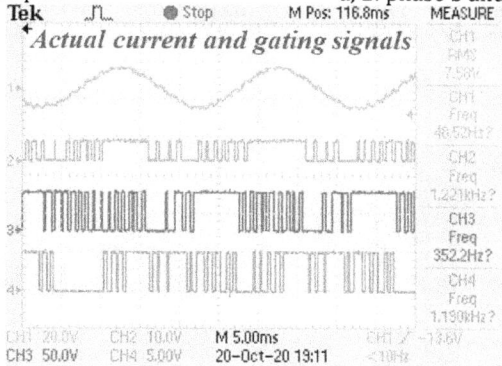

(g). 1: Actual current for phase a, 2: to 4: gating signals for top three switches of the inverter

Figure 4.5: Time response of the GFC for *Genmix 4*, delivering a power of 6 kW at 0.8pf with $\lambda=0.2$.

The other performance indices for this test case have been summarized in Table 4.4.

TABLE 4.4 PERFORMANCE INDICES OF MPC GRID-FORMING INVERTER WITH GFC

| Performance indices | IDM *Genmix* | | |
|---|---|---|---|
| | *Genmix 4* | *Genmix 4* | *Genmix 5* |
| Power references ($P^*$ and $Q^*$) | 6 kW at 0.8pf | 6 kW at 0.8 pf | 7 kW at 0.9 pf |
| $\lambda$ | 0.2 | 0.4 | 0 |
| $i^*(A)$ | 12.17 | 12.17 | 9.73 |
| $i(A)$ | 12 | 11.4 | 9.7 |
| $F_{asw}$ (kHz) | 1.1 | 0.87 | 1.23 |
| $THD_i$ (%) | 1.34 | 4.36 | 2.87 |
| LOH | 37 | 19 | 37 |
| $\varepsilon_i$ (%) | 1.40 | 6.33 | 0.3 |
| $u^*$ (V) | 239.6 | 239.6 | 239.6 |
| $u$ (V) | 237.1 | 228 | 238.1 |
| $\varepsilon_u$(%) | 1.0 | 4.8 | 0.6 |
| $THD_u$ (%) | 2.04 | 4.21 | 2.12 |

The GFC has been tested for another test condition where the IDM is in *Genmix 4* and delivering a power of 6 kW at 0.8 pf with λ=0.4. The test results obtained for this case are in Figures 4.6.(a) to (g). In this test case as well, the inverter has been successful of establishing a grid-forming mode of operation while tracking the power reference as required. With λ=0.4, the inverter current *THD* is increased slightly but with reduced $F_{asw}$ than *λ=0.2*. The current error can be visualized from Figure 4.6.(f). The other performance indices for this test case have been summarized in Table 4.4. The voltage and current *THD* values are observed to be 4.21 and 4.36 respectively. The *LOH* observed in this case is 19, which is much smaller than that of the previous case with λ=0.2. With increase in λ, the *LOH* has been shifted to low frequency side of the current spectrum, because of the reduction in the switching frequency.

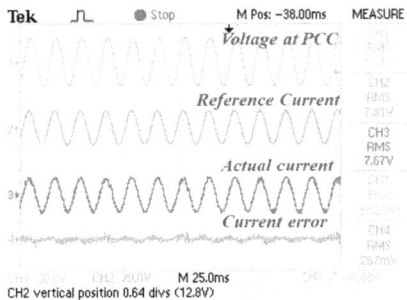

(a).1: Voltage at PCC, 2: reference current, 3: actual current and 4: instantaneous current error, *i*err.

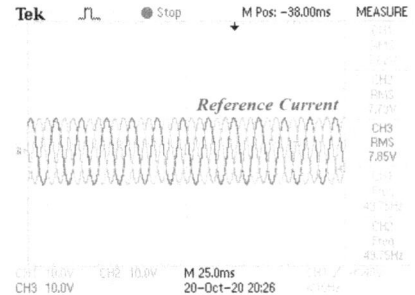

(d). Reference currents for phase a, phase b and phase c.

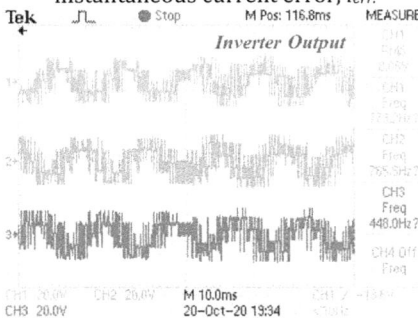

(b). Inverter output voltages- 1: Leg a, 2: Leg b and 3: Leg c.

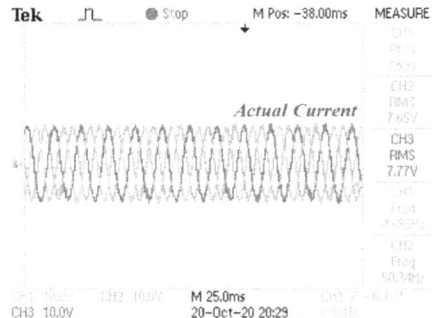

(e). Actual currents for phase a, phase b and phase c.

108

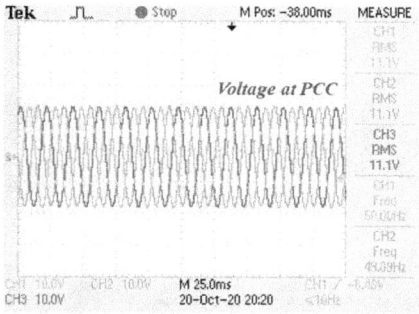

(c). Local grid voltages of phase a, phase b and phase c.

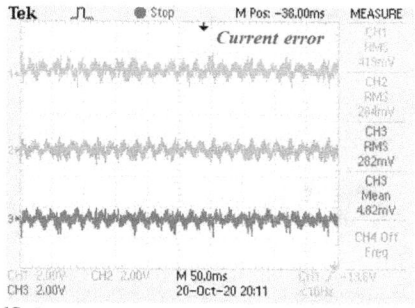

(f). Instantaneous current errors, 1: phase a, 2: phase b and 3: phase c.

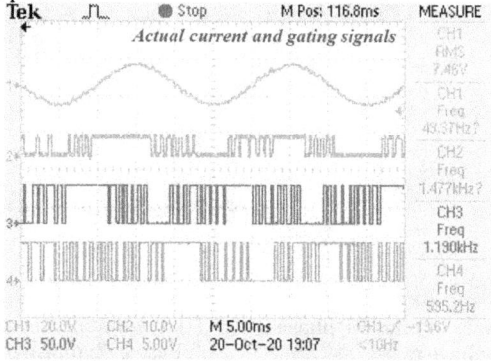

(g). 1: Actual current for phase a, 2: to 4: gating signals for top three switches of the inverter

Figure 4.6: Time response of the GFC for *Genmix 4*, delivering a power of 6 kW at 0.8 pf with $\lambda=0.4$.

Another test condition presented in the results of Figure 4.7 (a) to (e) corresponds to *Genmix 5* with a power reference of 7 kW at 0.9 pf with $\lambda=0$.

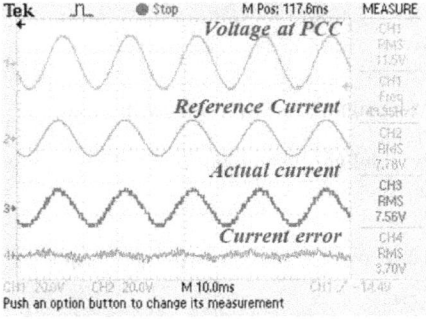

(a).1: Voltage at PCC, 2: reference current, 3: actual current and 4: instantaneous current error, $i_{err}$.

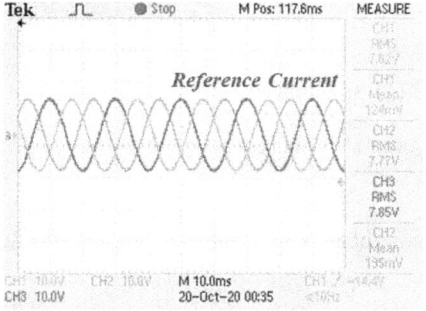

(d). Reference currents for phase a, phase b and phase c.

109

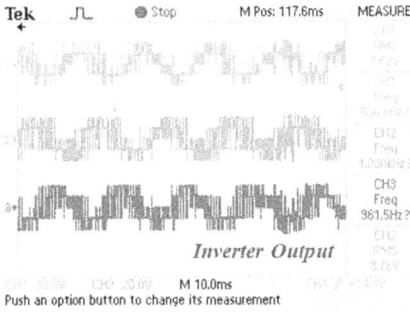

(b). Inverter output voltages- 1: Leg a, 2: Leg b and 3: Leg c.

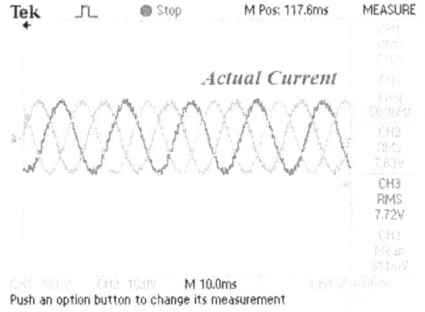

(e). Actual currents for phase a, phase b and phase c.

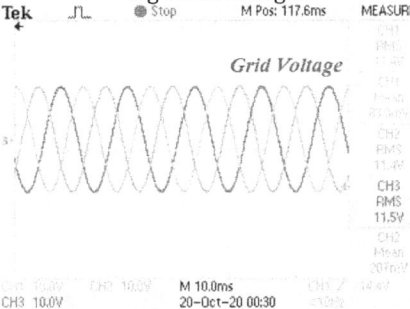

(c). Local grid voltages of phase a, phase b and phase c.

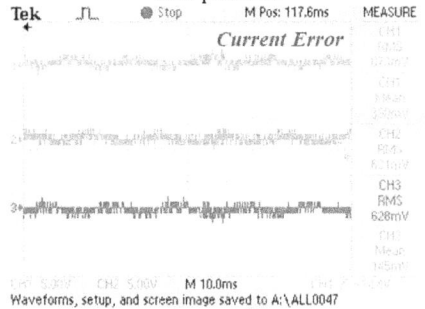

(f). Instantaneous current errors, 1: phase a, 2: phase b and 3: phase c.

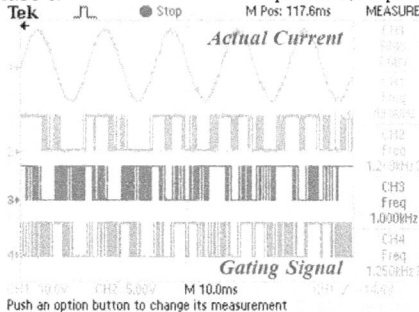

(g). 1: Actual current for phase a, 2: to 4: gating signals for top three switches of the inverter

Figure 4.7: Time response of the GFC for *Genmix 5*, delivering a power of 7 kW at 0.9 pf with $\lambda=0$.

Because of zero value taken for $\lambda$, the tracking is improved with high switching frequency.

## 4.2.4. Salient observations from the performance analysis of grid-forming control

- The developed GFC assist in the autonomous inverter operation following an intentional or unintentional islanding.
- The multi-objective weight function resulted in high efficiency and without considerable power tracking accuracy.

# 4.3 Multi-mode operation and mode transition of MPC inverters in IDM

As discussed in chapter 1, the inverters in IDM have to operate in grid-tied, grid-forming or grid-support mode as per the   requirements of IDM. Having tested the grid forming control, the transition across these modes is tested in the IDM considered for different operating scenarios.

## 4.3.1. Analysis of transition from grid-tied to grid-forming mode

Figures 4.8.(a) and (b) depict the developed GTC (Chapter 3) and GFC schemes respectively, and are reproduced here for ease of reference. The same 5-bus IDM considered in the previous section has been chosen as the test bench for the present study too. The inverter has been tested for all the *Genmix* configurations of Table 4.2 for mode transition control in the IDM considered through simulation as well as in the hardware test bench.

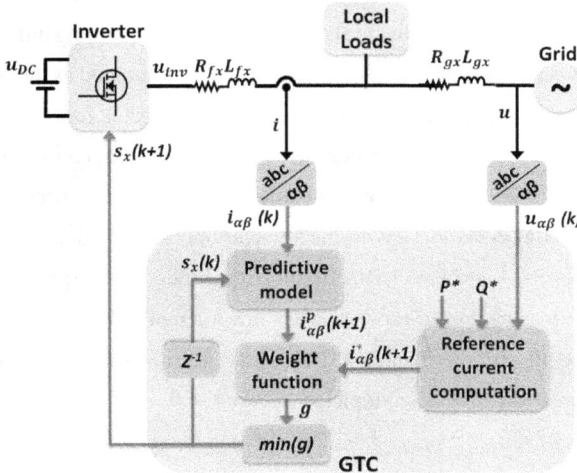

(a)

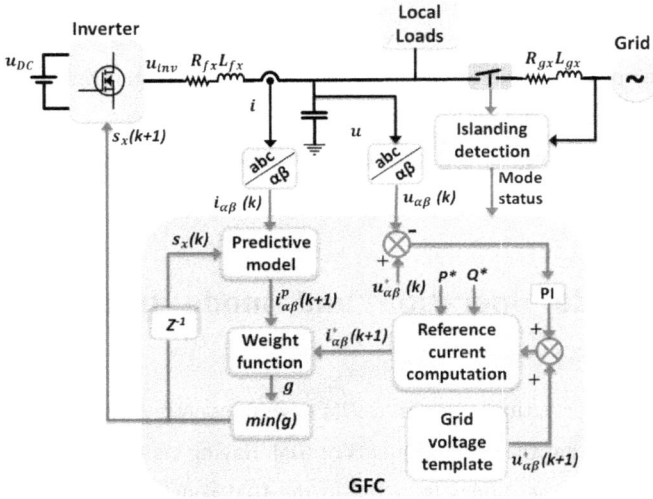

(b)
Figure 4.8: MPC inverter systems: (a) Grid-Tied Control (GTC), (b) Grid-Forming Control (GFC)

There are two types of mode transition defined for the IDM considered in this research work viz. (i) Inter mode transition, (ii) Intra mode transition. The intra mode transition will occur when the entire IDM is working in autonomous mode, and the operating conditions initiate a change in *Genmix*. For example, *Genmix 4* and *5* belong to this type. Upon islanding, if there is change in the RE generation condition may make the system to move from *Genmix 4* to *5* or vice versa. Here the inverter will be working in GFC mode just after islanding but may have to move to other *Genmix* in which it has to continue again in GFC. This is termed as intra mode transition. While working in GTC if any *Genmix* variation makes the inverter to move to GFC, then it is designated as inter mode transition. Simulation of mode transition control is implemented in MATLAB/Simulink and the results of one of the operating scenarios is presented below.

The former mode has IDM in *Genmix 1* with the inverter operating with GTC delivering a power of 6 kW at 0.8 pf with $\lambda=0.4$ and $F_s=50$ kHz. In the latter mode, the IDM configuration changed to *Genmix 4*, and the inverter has to move to grid-forming mode with GFC and support a local demand of 6 kW at 0.8 pf, with $\lambda=0.4$ and $F_s=50$ kHz. The reference and actual currents observed during this mode transition have been depicted in Figure 4.9.

112

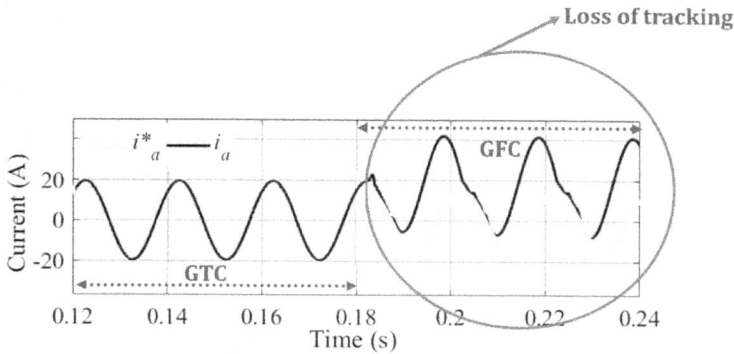
Figure 4.9: $i_a^*$ and $i_a$ during inter-mode transition from GTC to GFC.

As seen in Figure 4.9, loss of reference tracking is observed immediately after migration to *Genmix* 4 with GFC from *Genmix* 1 with GTC.

On analyzing the cause, it is found that the secondary function weightage in the MPC is the reason for the loss of tracking. In MPC, the average switching frequency of the inverter is dependent on the sampling frequency, $F_s$, as seen in chapter 3. But when the secondary function is added to reduce the switching frequency, a relative variation in inverter switching frequency occurs as commanded by the choice of $\lambda$. This made the primary target to be less prioritized compared to the switching frequency reduction. As a consequence of switching frequency reduction, the ripple in the inverter current deviated from the reference value especially when the system impedance is low in some *Genmix* condition, and lead to complete loss of tracking. As an example, migration from *Genmix* 1 to *Genmix* 4 working with same sampling frequency, the fixed $\lambda$ value of 0.4 for GFC in *Genmix* 4 is incapable of delivering the primary objective, whereas it met the tracking requirements in *Genmix* 1 with GTC. Thus, it is inferred that a fixed value of $\lambda$ that meets the performance requirements in a given mode of a particular *Genmix* does not suit other conditions in IDMs.

## 4.3.2. Test for the impact of $\lambda$ on reference tracking with GTC

Having identified the impact of $\lambda$ on the reference tracking with GFC controller with the multi objective weight function of equation (4.3), it is worth revisiting the tracking performance GTC the same weight function. A few *Genmix* conditions were considered to carry out the testing with varying $\lambda$ to study the tracking accuracy and the other performance indices defined earlier.

113

In this study, the GTC is targeted to deliver a $P*$ of 6 kW and a $Q*$ of 4.5 kVAR, which represent a reference current, $i^*_a$, of 20.01 A peak, while the IDM working in *Genmix 1*. The value of $\lambda$ is made to switch from 0 to 0.4 and then to 0.7 at periodic intervals when working with a fixed $F_s$ = 50 kHz. The result of this study is presented in Figure (4.10), in which, the reference tracking is accurate for values of $\lambda$ upto 0.4. When $\lambda$ reached 0.7 a complete loss of tracking is observed with unbound error, $i_{err}$, as evident from Figures 4.10 (a) & (b). Even the trajectory departs from the reference trajectory and goes out of bound.

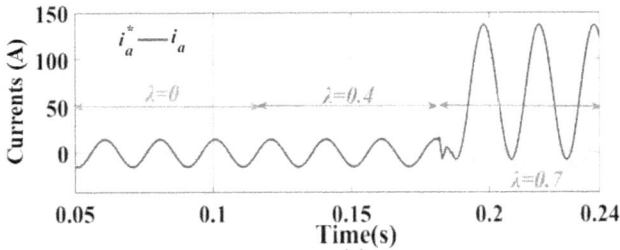

(a)

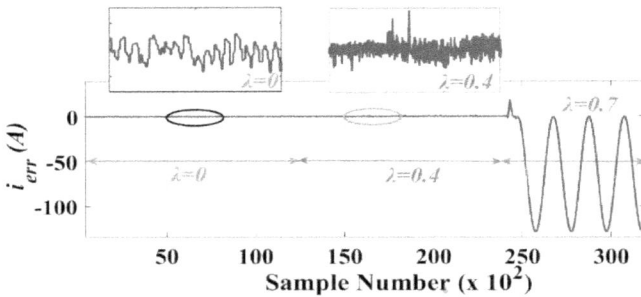

(b)

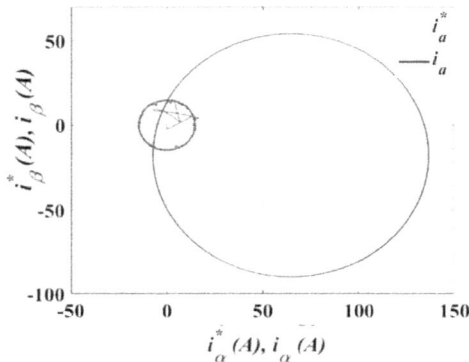

(c)

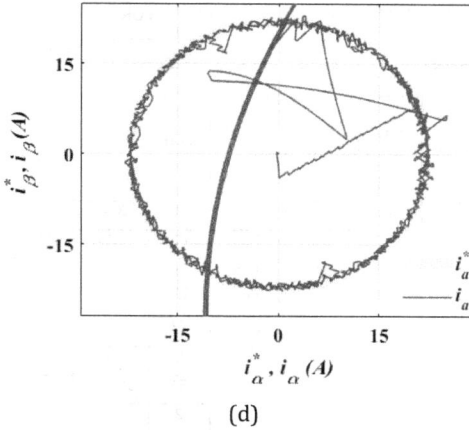

(d)

Figure 4.10: Time response of MPC inverter in grid-tied mode (a) $i^*_a$ and $i_a$; (b) Instantaneous current error, $i_{err}$; (c) & (d) Current vector trajectories of $i^*_a$ and $i_a$.

Once again it is inferred that the secondary function weightage has a significant impact on the tracking performance of any type of inverter controllers like GTC, GFC etc. Even in GTC one value of $\lambda$ which gave excellent performance is failing in other operating conditions of IDM. Therefore, it is decided to perform a sensitivity analysis on MPC, to ascertain the performance of inverter and deduce a correlation between the $F_s$, $\lambda$ and the grid impedance of each *Genmix* in any given IDM.

## 4.3.3. Sensitivity analysis of $F_s$ and $\lambda$ on the performance of MPC in inverters

The sensitivity analysis is carried on the MPC inverter of Figure 4.7 (a) through simulations in MATLAB/Simulink with the multi objective weight function of equation (4.4). The sampling frequency, $F_s$, in the model equation (4.3) is varied from 25 kHz to 100 kHz in steps of 25 kHz; $\lambda$ in equation (4.4) is also varied from 0 to 1 in steps of 0.1 at each $F_s$. This exercise is repeated for all the *Genmix* described Table 4.2 of IDM considered under study. Four performance indices from the earlier list have been identified for this sensitivity analysis, (i) average switching frequency, $F_{asw}$ in kHz, (ii) % $THD_i$, (iii) $LOH$, (iv) rms current error, $\varepsilon_i$.

TABLE 4.5 RESULTS OF SENSITIVITY ANALYSIS OF MPC INVERTER FOR DESIGN PARAMETER VARIATIONS

| $F_s$ in kHz | $\lambda$ | GT Mode & *Genmix 1* | | | | GF Mode | | | | | | | |
|---|---|---|---|---|---|---|---|---|---|---|---|---|---|
| | | | | | | *Genmix 4* | | | | *Genmix 5* | | | |
| | | $F_{asw}$ in kHz | THD in % | LOH | $\varepsilon_l$ in % | $F_{asw}$ in kHz | THD in % | LOH | $\varepsilon_l$ in % | $F_{asw}$ in kHz | THD in % | LOH | $\varepsilon_l$ in % |
| 25 | 0 | 22.14 | 3.58 | 94 | 3.66 | 23.19 | 1.51 | 88 | 3.55 | 17.32 | 0.96 | 53 | 3.31 |
| | 0.1 | 18.88 | 3.78 | 93 | 3.69 | 18.57 | 1.63 | 82 | 4.11 | 13.11 | 1.08 | 23 | 4.22 |
| | 0.2 | 16.58 | 3.98 | 85 | 3.88 | 14.87 | 1.74 | 64 | 4.63 | 9.53 | 1.3 | 30 | 5.15 |
| | 0.3 | 15.15 | 4.11 | 81 | 4.07 | 13.1 | 1.98 | 69 | 5.75 | 7.37 | 1.38 | 62 | 5.24 |
| | 0.4 | 14.28 | 4.18 | 80 | 4.11 | 11.09 | 2.09 | 53 | 6.27 | 6.015 | 1.57 | 23 | 5.28 |
| | 0.5 | 12.79 | 4.72 | 72 | 4.30 | 9.41 | 2.35 | 62 | 7.48 | 0.2 | 29.78 | 2 | 15 |
| | 0.6 | 12.64 | 4.76 | 71 | 4.54 | 7.127 | 2.6 | 51 | 8.65 | Non-Feasible Range | | | |
| | 0.7 | 11.17 | 4.89 | 70 | 7.74 | 0.61 | 17.74 | 2 | 22 | | | | |
| | 0.8 | 9.784 | 5.3 | 54 | 7.95 | Non-Feasible Range | | | | | | | |
| | 0.9 | 9.179 | 5.39 | 49 | 10.52 | | | | | | | | |
| | 1 | 8.694 | 5.76 | 45 | 10.75 | | | | | | | | |
| 50 | 0 | 43.74 | 1.73 | 162 | 1.26 | 45.75 | 0.8 | 183 | 0.94 | 35.12 | 0.48 | 181 | 1.38 |
| | 0.1 | 33.06 | 1.93 | 145 | 1.50 | 30.16 | 0.92 | 145 | 0.98 | 19.05 | 0.61 | 64 | 1.67 |
| | 0.2 | 28.17 | 2.02 | 146 | 1.59 | 22.34 | 1.13 | 113 | 2.34 | 12.63 | 0.8 | 39 | 4.33 |
| | 0.3 | 23.21 | 2.44 | 136 | 1.67 | 14.92 | 1.15 | 114 | 2.34 | 0.2 | 16.7 | 2 | 107.55 |
| | 0.4 | 20.99 | 2.72 | 114 | 5.00 | 0.74 | 18.48 | 2 | 22.00 | Non-Feasible Range | | | |
| | 0.5 | 17.84 | 2.89 | 80 | 5.10 | Non-Feasible Range | | | | | | | |
| | 0.6 | 13.05 | 3 | 14 | 9.91 | | | | | | | | |
| | 0.7 | 0.01 | 89.5 | 2 | 571 | | | | | | | | |
| 75 | 0 | 65.48 | 1.27 | 248 | 1.17 | 67.61 | 0.5 | 271 | 0.16 | 52.6 | 0.32 | 119 | 0.43 |
| | 0.1 | 44.15 | 1.28 | 216 | 1.24 | 37.95 | 0.66 | 195 | 0.20 | 23.51 | 0.47 | 106 | 0.25 |
| | 0.2 | 36.06 | 1.62 | 180 | 1.25 | 22.55 | 1 | 21 | 0.38 | 0.2 | 21.7 | 2 | 107.55 |
| | 0.3 | 27.8 | 1.72 | 173 | 4.19 | 0.82 | 33.01 | 2 | 22.00 | Non-Feasible Range | | | |
| | 0.4 | 21.9 | 1.88 | 167 | 5.94 | Non-Feasible Range | | | | | | | |
| | 0.5 | 0.01 | 89.5 | 2 | 570 | | | | | | | | |
| 100 | 0 | 87.73 | 0.96 | 342 | 0.94 | 91.08 | 0.37 | 355 | 1.03 | 69.91 | 0.25 | 193 | 0.91 |
| | 0.1 | 55.71 | 1.08 | 302 | 1.68 | 43.45 | 0.54 | 232 | 1.96 | 25.57 | 0.65 | 82 | 1.29 |
| | 0.2 | 40.75 | 1.33 | 196 | 2.90 | 1.22 | 20.84 | 2 | 22.00 | 0.2 | 20.83 | 2 | 108.00 |
| | 0.3 | 25.42 | 1.72 | 9 | 12.16 | Non-Feasible Range | | | | Non-Feasible Range | | | |
| | 0.4 | 0.01 | 89.5 | 2 | 570.0 | | | | | | | | |

### 4.3.4. Salient observations from the sensitivity analysis

From the sensitivity analysis the following observations have been made on the performance of MPC inverter:

- Under all conditions, the average switching frequency, $F_{asw}$, is observed to decrease with increase in $\lambda$ for all values of sampling frequency.

- Generally, *LOH* are found to be high enough, indicating small distortion factors. But, *LOH* are found to decrease with increase in $\lambda$ and can be correlated with the above observation.

- For obtaining such high *LOH* values as that of Table 4.5 with a typical sine PWM inverter, if used in the same application, it has to operate at double the switching frequency as that of the present $F_{asw}$, which will double the switching losses. This characteristic proves the competency of multi-objective MPC.

- The current error, $i_{err}$, and $THD_i$ have been observed to decrease with decrease in $\lambda$. This feature can be utilized to improve the harmonic performance of the MPC whenever needed.

- 'Non-Feasible Range' in Table 4.5 represents the range of $\lambda$ for which loss of tracking is observed. Increase in either sampling frequency or the system impedance or both shall decrease the limiting value of $\lambda$. When the inverter is operated with low sampling frequency in a low impedance network, $\lambda$ did not exhibit a non-feasible range.

- The basic MPC with a rigid choice of $\lambda$, when deployed in IDMs, may result in reference tracking failure if the combination of $\lambda$ and $F_s$ falls into the non-feasible range owing to changes in *Genmix* of IDM.

The results of this study, reveals that when designing MPC for IDMs, a right combination of the design parameters has to be identified for every operating modes and operating conditions to avoid system failure. This necessitates the formulation of a strategy that can ensure power tracking by avoiding the operation in the non-feasible range to avoid reference tracking failure in the inverter. Therefore, the design parameters of MPC inverter must be assigned dynamically based on the operating conditions in IDMs. This realization following the sensitivity analysis has directed the present study towards the next step that aims to impart a self-adaptability feature to MPC inverters in IDM. Thus, Table 4.5 serves as the database for imparting dynamic adaptability feature in MPC controllers.

## 4.4. Dynamic adaptability for MPC grid-tied inverters

Dynamic Adaptability (DA) is a technical term coined in this thesis that reflects self-adaptability in multi-mode operation and mode transition of MPC based grid-tied inverters exclusively for IDM. Infusing such self-adaptability in inverter control has been targeted as the major focus in the upcoming sections.

### 4.4.1. Dynamic adaptability

DA enables the conventional MPC scheme to suitably redefine its design parameters to match the requirements of dynamic systems like IDM. DA can attribute two significant characteristics to the MPC inverter defined below.

**(i). Multi-mode operation and mode transition**

- This feature enables the MPC inverters to operate in any of the two modes and shift from grid-tied mode to grid-forming mode or vice versa as per demand. The design parameters are selected in such a way that in each mode the loss of tracing is avoided. This can ensure stable operation in every mode under all possible operating conditions of the IDM considered.

**(ii). Dynamic parameter assignment**

- This feature helps to dynamically update the static system model and design parameters of MPC in order to follow the system dynamics and accomplish the desired performance in each mode of operation.

### 4.4.2. Dissemination approach for dynamic adaptability database

The dynamic adaptability database is embedded in the MPC controller as a look-up table and is accessed through three priority signals of the controller settings $\rho_H$, $\rho_L$ and $\rho_T$ are the priority signals to meet harmonic standards, inverter losses and reference tracking, respectively. These priority signals are to be set either as high or low as demanded by the micro-grid operator. The following rules have been formulated to meet the priorities.

- When no priority is set, then the mid value in the feasible range of $\lambda$ will be picked by the controller.
- When any one or more of the priorities is/are set high, then value of $\lambda$ will be decided as stated below:

(i)     If $\rho_H$ is enabled high, it demands to prioritize the harmonic levels of the inverter output; so, the minimum possible value of $\lambda$, even as low as 0, will be chosen.

(ii)    If $\rho_L$ is enabled high, it demands reduction in inverter losses; so, the maximum possible value of $\lambda$ will be chosen within its feasible range.

(iii)   If $\rho_T$ is enabled high, it urges on a perfect reference tracking and $P^*$ delivery; so, a value of $\lambda$ below its mid value in the feasible range will be picked.

(iv)    If both $\rho_T$ and $\rho_L$ are enabled high, then it demands both reference tracking and loss minimization simultaneously; so, the mid value of $\lambda$ in the feasible range will be picked.

Similarly, all possible combinations of the priorities are assigned as a rule base to pick one unique value of $\lambda$ for any given operating condition.

The dynamic adaptability developed has been imparted to the MPC inverter and then validated through simulation studies and hardware experiments.

## 4.5 Dynamically adaptable model predictive control

A Dynamically Adaptable Model Predictive Control (DAMPC) scheme has been developed and presented in this thesis. The developed DAMPC scheme is shown in Figure 4.11.

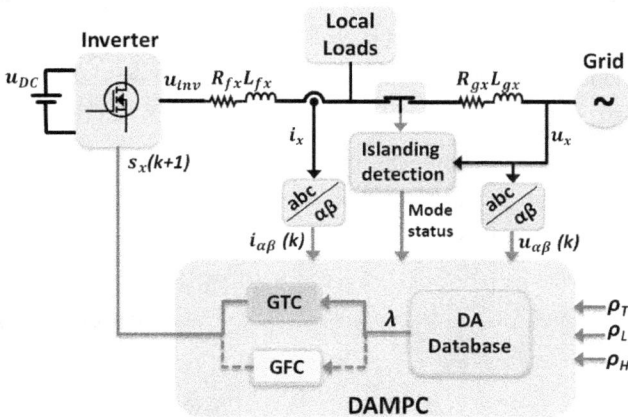

Figure 4.11: Dynamically Adaptable Model Predictive Control, DAMPC.

It is targeted to assess the 'multimode operation and mode transition' ability in various IDM operating scenarios such as,

(i)     Inter-mode Operation: - switching from grid-tied mode to any *Genmix* of grid-forming mode, and,

(ii)    Intra-mode Operation: - switching from one *Genmix* to another within grid-forming mode.

DAMPC also enables these multi-mode transitions with dynamic assignment of $\lambda$ and sampling frequency. The steps followed for development of the DAMPC has been illustrated in Figure 4.12.

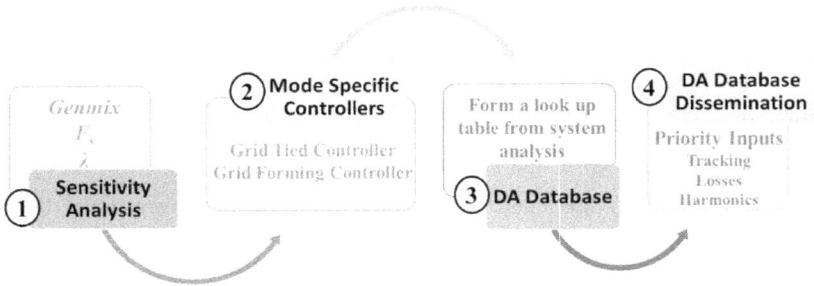

Figure 4.12: Development phases of DAMPC.

## 4.5.1. Testing of DAMPC in simulated IDM

The same 5-bus IDM test bench shown in Figure 4.2 has been chosen for validation of DAMPC. The DAMPC is implemented in the solar plant of the IDM considered and tested for both inter-mode and intra-mode transitions with various combinations of *Genmix* and priority signals.

**(a) Test 1 – Inter-mode operation:**

The DAMPC inverter is initially made to work in grid-tied mode with *Genmix 1* to deliver a P* of 6 kW and Q* of 4.5 kVAR without any priority signal enabled. The grid frequency is assumed as 49.95 Hz and the voltage regulation target of the IDM is kept at ±1%. The DA database recommended a $\lambda$ of 0.4 for *Genmix 1*. The inverter gets an island detect signal at the instant of 1s at which the inverter has to move to grid-forming mode in *Genmix 5*. The DAMPC suggested $\lambda$ =0.3 for grid-forming mode as it is the mid value of the feasible range of $\lambda$. The two mode specific controllers, GTC and GFC, have been designed to work with the weight function of equation (4.4) and a sampling frequency of 50 kHz.

The instantaneous quantities of the DAMPC inverter operation for inter-mode operation are shown in Figures 4.13.(a) to (h), where, $i^*_a$, is the reference current in grid-tied mode and $i^*_{a-gf}$, is that in grid-forming mode. The peak reference currents before and after islanding are seen as 20.01 A and 15 A respectively for the commanded P*. The inverter current is observed to track the commanded references in the respective modes.

Upon the mode change, the controller is found tracking the new reference signal within a transition time of about 1.5 ms, during which it establishes rated voltage and frequency in the local grid as seen in Figures 4.13.(f) & (g).

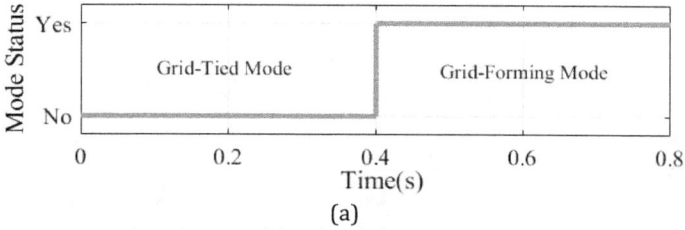

(a)

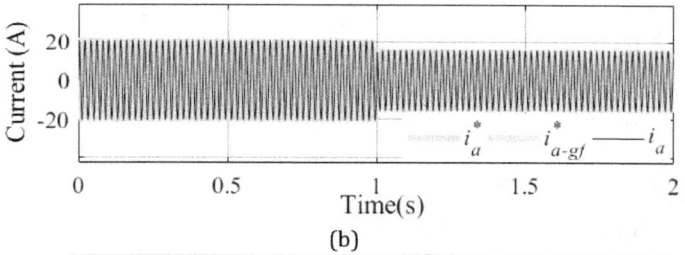

(b)

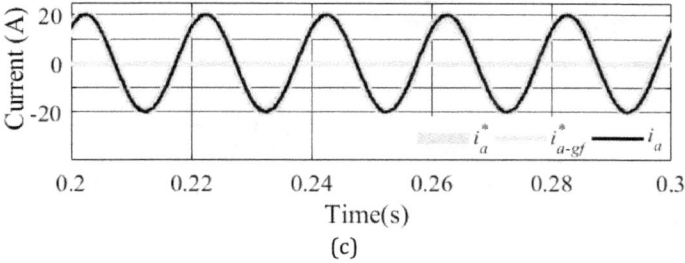

(c)

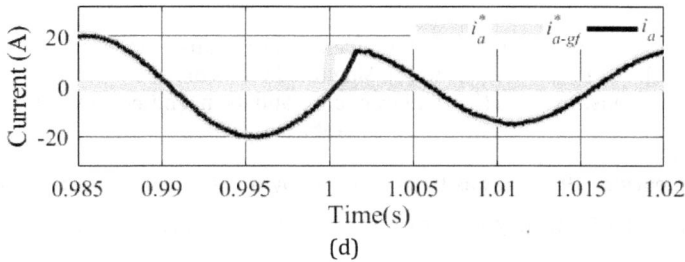

(d)

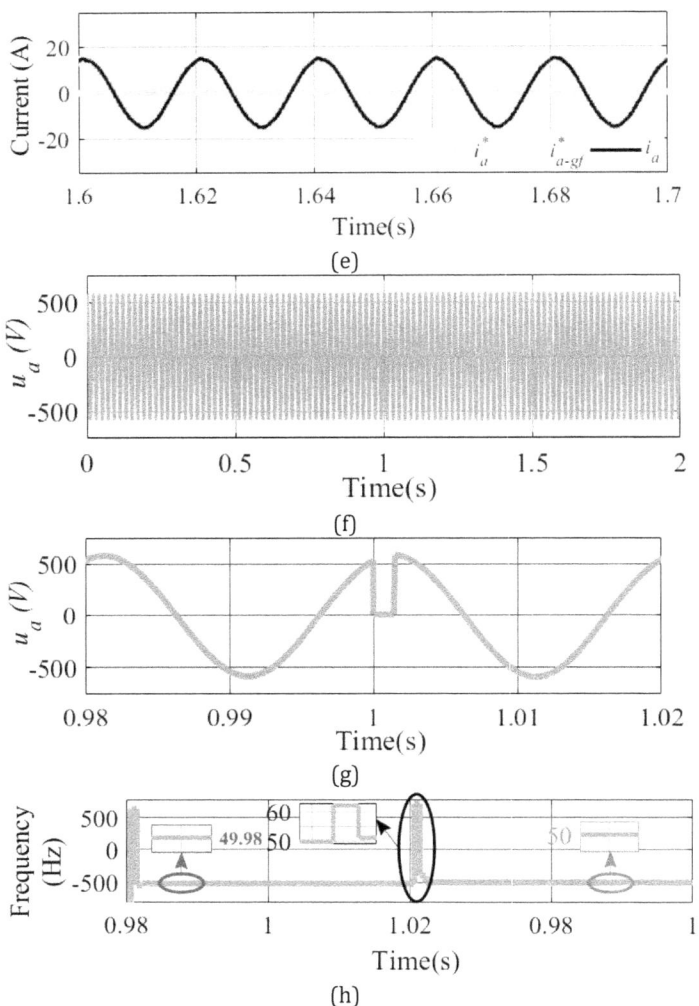

Figure 4.13: Time response of the DAMPC inverter for the inter-mode transition: (a): Mode status signal; (b): $i*_a$, $i*_{a\text{-}gf}$ and $i_a$; (c) to (e): Magnified views of $i*_a$, $i*_{a\text{-}gf}$ and $i_a$ (f) and (g): $u_x$ at the instant of transition, and (h): Grid frequency and its magnified views three different instants.

The grid frequency of 49.98 Hz has been followed by the inverter current in grid-tied mode, whereas the inverter frequency is at its rated value after the mode transition as seen in Figure 4.13.(h).

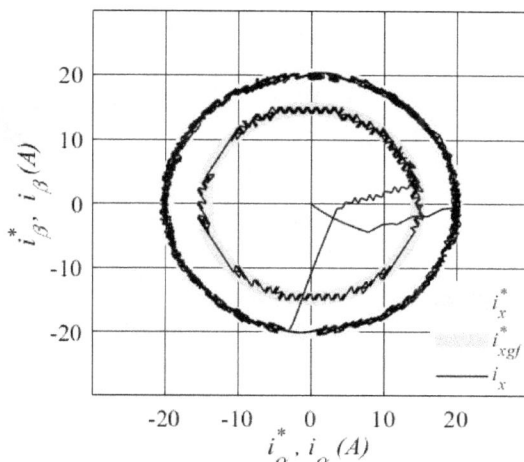

Figure 4.14: Current vector trajectory of $i^*_x$, $i^*_{x\text{-}gf}$ and $i_x$ of the DAMPC inverter for the inter-mode transition

The current vector trajectory for this transition with two circular trajectories each representing one mode is shown in Figure 4.14. Both the trajectories are well within the reference trajectories confirming the accuracy of tracking before, during and after mode change.

### (b) Test 2: Intra-mode operation

The DAMPC inverter has been tested for intra-mode transitions with GFC only. The former and latter modes descriptions are as follows:

- Initially the inverter is working in *Genmix 5* delivering a P* of 8 kW at 0.8 pf lag and operated at sampling frequency 50 kHz.
- The IDM is then switched to *Genmix 4* at the instant of 1s to deliver a new P* of 6 kW at 0.8 pf lag and at the same sampling frequency.
- The operator priority signal, $\rho_L$, was enabled in both the cases of Genmix.

The DAMPC feature suggested $\lambda$=0.3 for *Genmix 5* and $\lambda$=0.4 for *Genmix 4*, as the operator priority demanded minimum inverter losses; the suggested values of $\lambda$ are the maximum values in the respective feasible ranges. Figures 4.15.(a) and (b) present the time window during the mode transition from *Genmix 5* to *Genmix 4* depicting an intra-mode transition time of 2 ms.

123

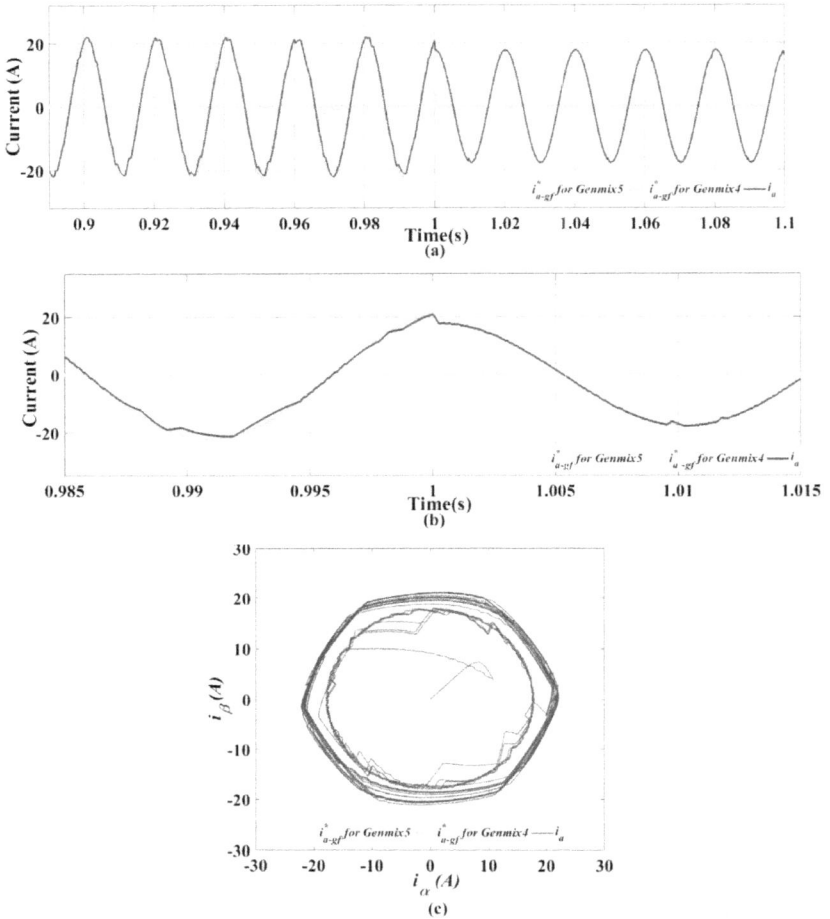

Figure 4.15: Intra-mode transition performance of DAMPC: (a): $i^*_{a\text{-}gf}$ for *Genmix 5*, $i^*_{a\text{-}gf}$ for *Genmix 4* and $i_a$, (b): Magnified view of $i^*_{a\text{-}gf}$ for *Genmix 5*, $i^*_{a\text{-}gf}$ for *Genmix 4* and $i_a$ at the instant of intra-mode transition, and, (c): Current vector trajectories of $i^*_{a\text{-}gf}$ for *Genmix 5*, $i^*_{a\text{-}gf}$ for *Genmix 4* and $i_a$.

The inverter current peaks are found in Figure 4.15.(b) as 20.01 A and 18.75 A in *Genmix 5* and *4* respectively, which are in good agreement with the respective $P^*$ values. The trajectories of the current vectors in Figure 4.15.(c) show high level of tracking accuracy with respect to the corresponding reference current trajectories.

The performance indices of the DAMPC inverter obtained from the simulation study has been summarized in Table. 4.6.

| Mode transition | Inter-mode transition | | Intra-mode transition | |
|---|---|---|---|---|
| **IDM condition** | Former mode *Genmix 1 & grid-tied* | Latter mode *Genmix 5 & grid-forming* | Former mode *Genmix 5 & grid-forming* | Latter mode *Genmix 4 & grid-forming* |
| $\lambda$ | 0.1 | 0.3 | 0 | 0.2 |
| $F_s$ (kHz) | 25 | 25 | 25 | 25 |
| $F_{asw}$ (kHz) | 1.8 | 1.31 | 1.732 | 0.953 |
| $THD_i$ (%) | 3.78 | 1.98 | 0.96 | 1.3 |
| LOH | 93 | 69 | 53 | 30 |
| $\varepsilon_i$ (%) | 3.69 | 4.2 | 0.3 | 1.1 |

Across the entire simulation analysis, the DAMPC inverter has been found to successfully accomplishing multi-mode transition without any failure in tracking in every mode and in *Genmix*.

## 4.5.2. Hardware validation of DAMPC in IDM

The laboratory testing and validation of DAMPC inverter have been conducted with RT-Lab and OP4200 Hardware-In-Loop Simulator (OPAL-RT). The solar PV inverter in the laboratory scale IDM emulator is controlled with the developed DAMPC for various Genmix conditions while the DAMPC is programmed in OPAL-RT. The snapshot of the experimental setup of the IDM and the controller are shown in Figure 4.16.

A 10 kVA three phase IGBT inverter of the solar PV plant with specifications as in Table 4.1 is synchronized to the IDM emulator. The system and component specifications are the same as in Table 4.2. The DAMPC has been tested for its mode transfer capability with desired performance as demanded by the priority signals for various operating modes of IDM. The same set of performance indices used in the earlier testing were followed here and are reproduced below for quick reference.:

(i)     average switching frequency, $F_{asw}$ in kHz

(ii)    inverter current harmonics, $THD_i$ in %

(iii)   lowest order harmonics, *LOH*

(iv)   rms current error, $\varepsilon_i$, in %

(v)    rms voltage error, $\varepsilon_u$, in %

(vi)   voltage harmonics, $THD_u$ in % (only in GFC)

(vii)   mode transition time in sample and cycle level in s

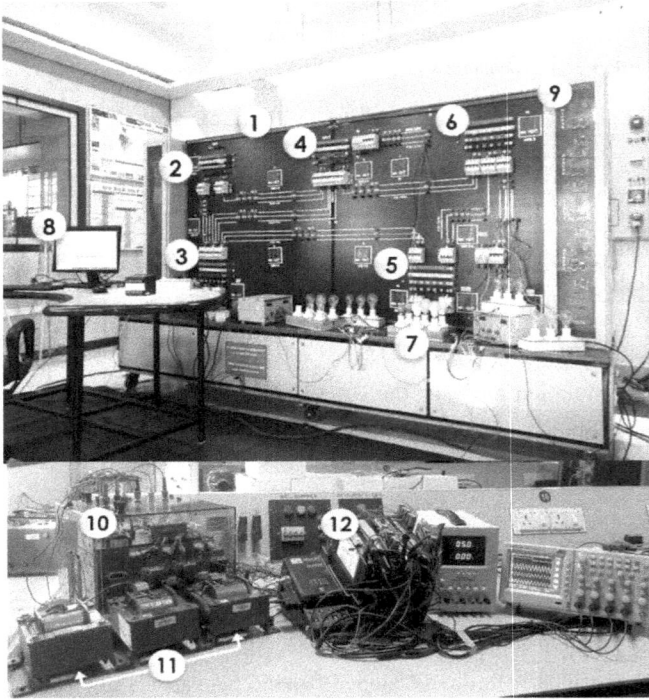

Figure 4.16: IDM test bench and the experimental setup of DAMPC inverter: 1. IDM emulator, 2. grid connection point, 3. wind turbine generator Bus, 4. solar PV Bus, 5. hydroelectric generator Bus ,6. radial feeder bus, 7. Local loads, 8.  HIL desktop, 9. Communication network terminals ,10.  Three phase IGBT inverter (Semikron), 11. Filter, 12. OP4200 HIL simulator.

Both inter-mode and intra-mode transitions have been initiated with appropriate *Genmix* conditions of IDM and the inverter performance has been analyzed and presented.

## 4.5.3. Experimental results

All the waveforms are captured at the mode transition edge using single shot mode in the DSO, with all voltages presented in 20× scale and currents in 2× scale.

**(a) inter-mode transitions:**

- **Grid-tied to Grid-forming**

Case (i). While working with GTC, the IDM condition is changed so that the inverter control is expected to move to GFC with the specifications as described in the Table.4.7 below.

| IDM condition | Former mode: grid-tied with *Genmix 1* | Latter mode: grid-forming with *Genmix 5* |
|---|---|---|
| Power reference | 6kW at 0.8 pf | 6kW at 0.8 pf |
| Priority inputs | $\rho_H$ and $\rho_L$ are enabled | $\rho_L$ is enabled |
| DA suggested $\lambda$ | 0.1 | 0.3 |

Figures 4.17. depicts the results of this inter-mode transition, wherein, GFC is found maintaining a stable grid voltage and frequency, while the inverter current magnitudes strictly adhere to the demanded power. With $\lambda$=0.3, in the latter mode, the reduction in switching frequency is evident as seen in Figure 4.17.(b). In both the modes the inverter currents follow the reference currents as depicted in Figures 4.17. (d) and (e). From Figure 4.17.(f), a slightly increased current distortions are seen in the latter mode due to the higher value of $\lambda$.

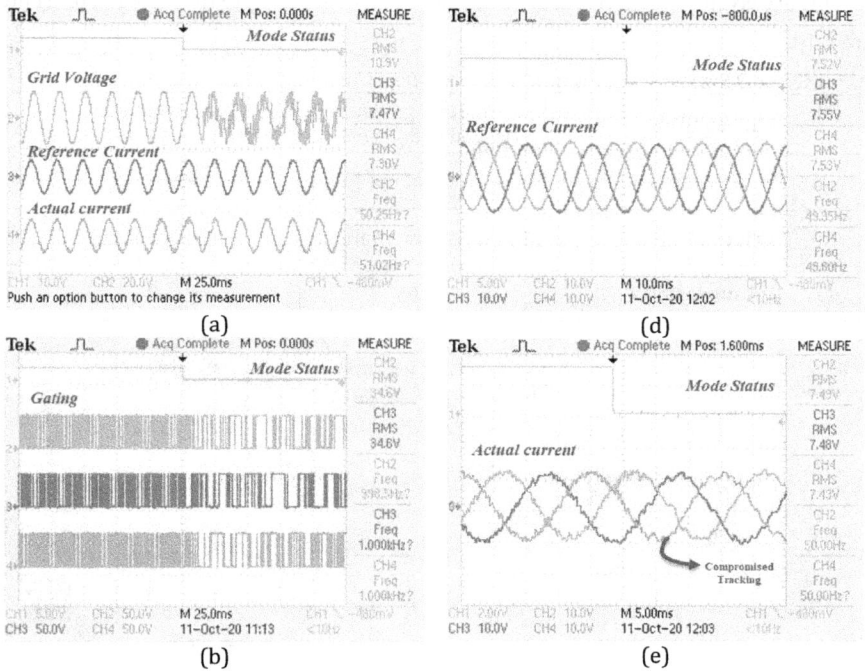

(a)

(b)

(d)

(e)

127

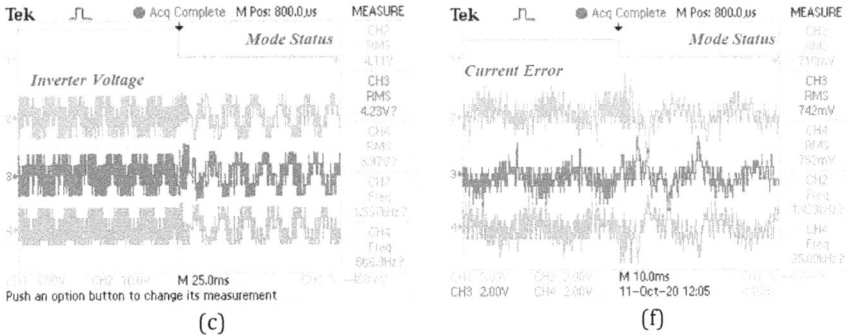

(c)                                                                              (f)

Figure 4.17: Inter-mode transition from GTC to GFC: (a). grid voltage and currents, (b). Gating signals for top three inverter switches the; (c). inverter leg voltages; (d). reference currents; (e). actual currents; (f) current errors.

Case (ii). Another case of inter-mode transition is considered with GTC to GFC but with two different power references in two modes. The specifications of the former and latter mode conditions are given in the table below. The results corresponding to this case are presented in Figure 4.18.

TABLE.4.8 SYSTEM SPECIFICATIONS FOR GRID-TIED TO GRID-FORMING TRANSITION CASE (ii)

| IDM condition | Former mode:<br>grid-tied with *Genmix 1* | Latter mode:<br>grid-forming with *Genmix 4* |
|---|---|---|
| Power reference | 5 kW at upf | 7 kW at 0.8pf |
| Priority inputs | $\rho_L$ and $\rho_T$ are enabled | $\rho_T$ is enabled |
| DA suggested λ | 0.4 | 0 |

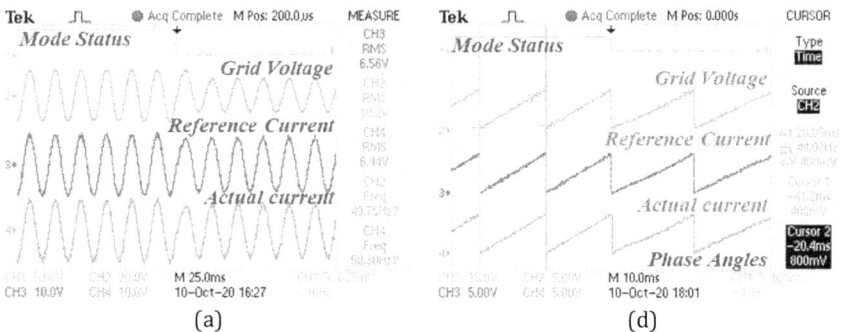

(a)                                                                              (d)

128

(b)

(e)

(c)

(f)

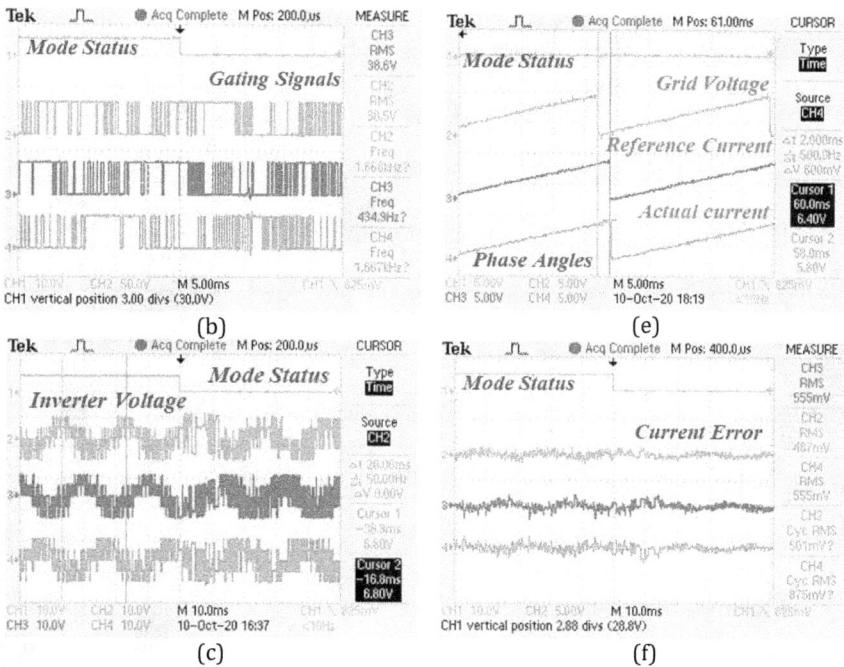

Figure 4.18: Inter-mode transition from GTC to GFC: (a). Grid voltage and currents (b). Gating signals for top three inverter switches (c). inverter leg voltages; (d) and (e). phase angles of grid voltage, reference current and actual current (f) three phase current errors.

All the results exactly follow the same trend as in the previous cases. But a key observation is arrived from the reference and the actual currents of Figure 4.18 (a), that before and after the transition the respective values are computed and followed by the controllers. This ensured the power balance in the IDM upon migration from any *Genmix* to the other. Another salient observation is the priorities, a $\lambda$ of 0.4 is chosen in the former mode, because of both the inverter loss and tracking priorities, while $\lambda = 0$ is chosen in the latter mode. So, the current error in the latter mode is found to be extremely low, with high switching frequency from Figures 4.18 (b) & (f).

More test cases were considered and their results are presented in Table 4.9, which summarizes all the inverter performance indices corresponding to all the test condition considered in the IDM.

TABLE.4.9 HARDWARE TEST RESULTS

| Performance indices | Inter-mode transition: Grid-Tied to Grid Forming | | | | | |
|---|---|---|---|---|---|---|
| | Case (i) | | Case (ii) | | Case(iii) | |
| | Former mode: grid-tied with *Genmix 1* | Latter mode: grid-forming with *Genmix 5* | Former mode: grid-tied with *Genmix 1* | Latter mode: grid-forming with *Genmix 4* | Former mode: grid-tied with *Genmix 1* | Latter mode: grid-forming with *Genmix 5* |
| Power reference | 6kW at 0.8 pf | 6kW at 0.8 pf | 5 kW at upf | 7 kW at 0.8pf | 6 kW at 0.9pf | 6 kW at upf |
| Priority inputs | $\rho_H$ and $\rho_L$ | $\rho_L$ | $\rho_L$ and $\rho_T$ | $\rho_T$ | $\rho_H$ | $\rho_L$ |
| DA suggested $\lambda$ | 0.1 | 0.3 | 0.4 | 0 | 0 | 0.4 |
| $F_{asw}$ (kHz) | 1.24 | 0.94 | 0.72 | 1.25 | 1.26 | 1.06 |
| THD (%) | 4.35 | 4.85 | 3.2 | 2.13 | 2.14 | 2.01 |
| LOH | 38 | 22 | 38 | 50 | 35 | 43 |
| $\varepsilon_i$ (%) | 4.12 | 5.47 | 0.33 | 0.09 | 2.12 | 0.41 |
| $\varepsilon_v$ (%) | 0.0 | 2.9 | 0.0 | 0.9 | 0.7 | 0.3 |
| $THD_v$ (%) | 0.1 | 4.92 | 2 | 0.1 | 2.46 | 1.87 |

- **Grid-forming to grid-tied mode**

Two cases of inter-mode transition from grid-forming to grid-tied mode are discussed in this section.

Case (i). The former mode and latter mode conditions as given in the Table 4.10 below, and the results of this transition are presented in Figures 4.19.

TABLE.4.10 SYSTEM SPECIFICATIONS FOR GRID-FORMING TO GRID-TIED MODE TRANSITION CASE (i)

| IDM condition | Former mode: grid-forming with *Genmix 5* | Latter mode: grid-tied with *Genmix 1* |
|---|---|---|
| **Power reference** | 7kW at upf | 5kW at 0.8 pf |
| **Priority inputs** | $\rho_L$ is enabled | $\rho_H$ is enabled |
| **DA suggested $\lambda$** | 0.3 | 0 |

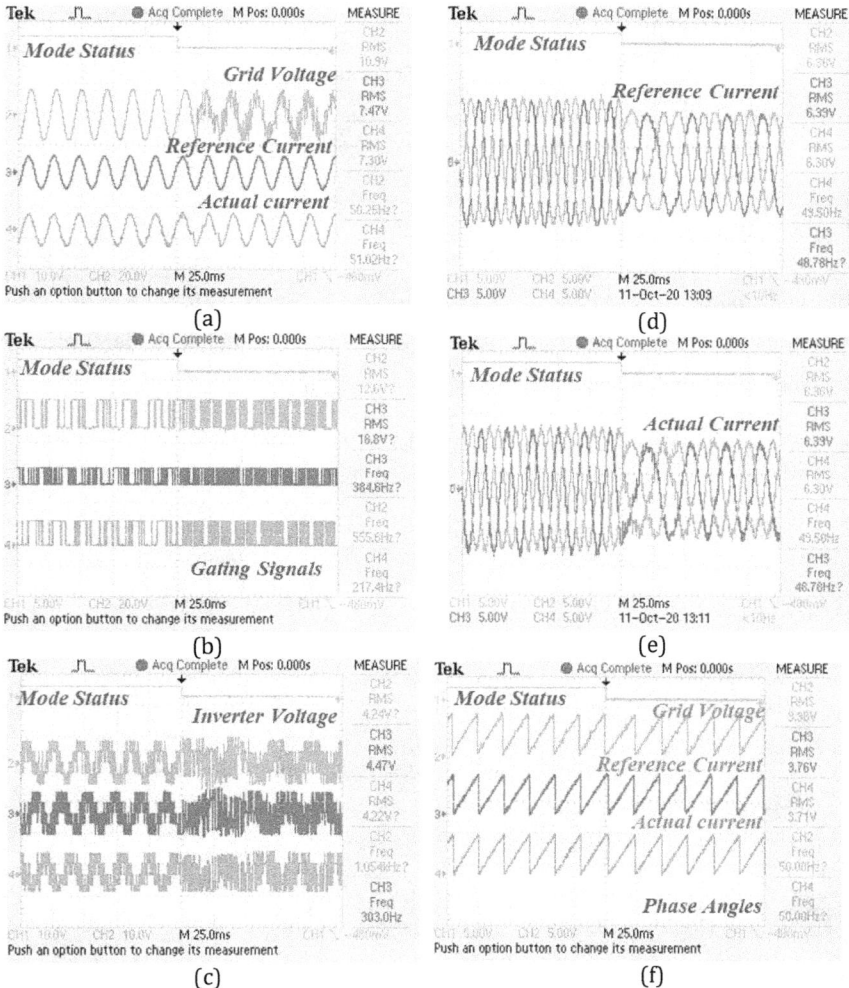

Figure 4.19: Inter-mode transition from GFC to GTC: (a). Grid voltage and currents, (b). Gating signals for top three inverter switches; (c). inverter leg voltages; (d) reference current; (e) actual current and (f). phase angles

The grid voltage, the gating signals and the inverter output voltage shown in different Figures ascertains that GFC established the grid successfully, and reduction in switching frequency because of the enabled $\rho_L$.

Case (ii). The second case of inter-mode transition from grid-forming to grid-tied mode is shown in Figures 4.20.(a) to (f). The former mode and latter mode conditions are:

| IDM condition | Former mode: grid-forming with *Genmix 5* | Latter mode: grid-tied with *Genmix 1* |
|---|---|---|
| Power reference | 6 kW at 0.8pf | 6 kW at 0.95pf |
| Priority inputs | $\rho_H$ is enabled | $\rho_L$ is enabled |
| DA suggested $\lambda$ | 0.2 | 0.6 |

The observed quantities for this test case are presented in Figure 4.20. Even in this case, the reference tracking is found accurate with the switching frequency and the current error following the same trend against $\lambda$.

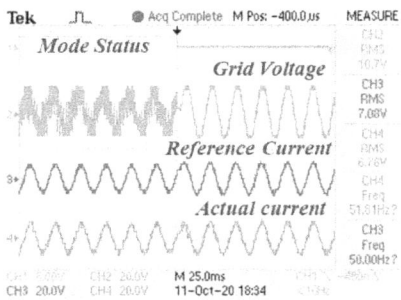

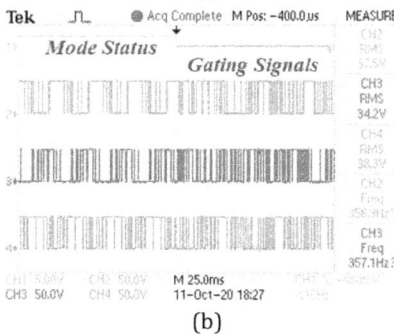

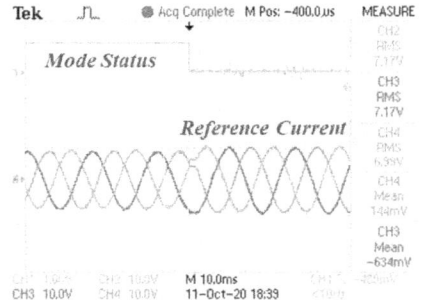

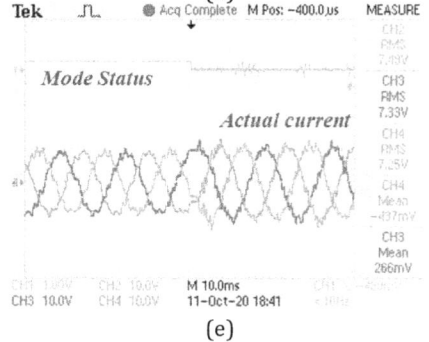

(a)

(b)

(d)

(e)

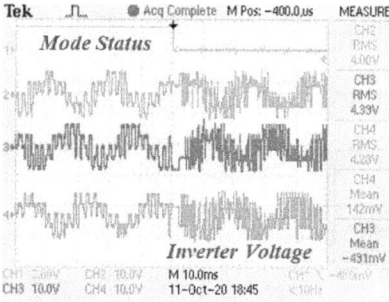

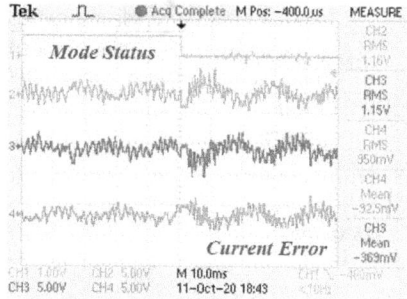

(c)                                              (f)

Figure 4.20: Inter-mode transition from grid-forming to grid-tied mode: (a). Grid voltage and currents, (b). Gating signals for top three inverter switches (c). inverter leg voltages; (d) reference current; (e) actual current; (f) current errors.

Table 4.12 summarizes all the performance indices of the inverter under all the inter-mode transitions cases of GFC to GTC.

TABLE.4.12 HARDWARE TEST RESULTS

| Performance indices | Inter-mode transition: grid-forming to grid-tied | | | |
|---|---|---|---|---|
| | Case (i) | | Case (ii) | |
| | Former mode:grid-forming with Genmix 5 | Latter mode: grid-tied with Genmix 1 | Former mode: grid-forming with Genmix 5 | Latter mode: grid-tied with Genmix 1 |
| Power reference | 7kW at upf | 5kW at 0.8 pf | 6 kW at 0.8pf | 6 kW at 0.95pf |
| Priority inputs | $\rho_L$ | $\rho_H$ | $\rho_H$ | $\rho_L$ |
| DA suggested $\lambda$ | 0.3 | 0 | 0.2 | 0.6 |
| $F_{asw}$ (kHz) | 0.875 | 1.23 | 1.12 | 0.47 |
| THD (%) | 4.26 | 2.87 | 2.57 | 5.23 |
| LOH | 23 | 37 | 25 | 17 |
| $\varepsilon_i$ (%) | 0.8 | 0.3 | 0.48 | 6.61 |
| $\varepsilon_v$(%) | 0.0 | 0.6 | 1.1 | 0.0 |
| $THD_v$ (%) | 3.24 | 2.12 | 3.14 | 5.12 |

**(b) Intra-mode transitions:**

- **Genmix 5 to Genmix 4**

Three cases of Intra-mode transition, all within grid-forming mode due to *Genmix* variation in IDM, are discussed in this section.

Case (i). In the first case, the former mode and latter mode conditions are:

133

| IDM condition | Former mode:<br>grid-forming with *Genmix 5* | Latter mode:<br>grid-forming with *Genmix 4* |
|---|---|---|
| Power reference | 7 kW at upf | 9 kW at upf |
| Priority inputs | $\rho_T$ is enabled | $\rho_H$ is enabled |
| DA suggested λ | 0 | 0.2 |

The results corresponding to this transition from *Genmix* 5 to 4 are shown in Figures 4.21. The tracking error is found as 0.41% and 2.12% respectively in the former and latter modes as the former is bound to give utmost accuracy because of the enabled $\rho_T$. The other trends like the magnitude of the delivered current following the power command variation, switching frequency's inverse relation with λ, and current error's direct relation with λ were all seen even here.

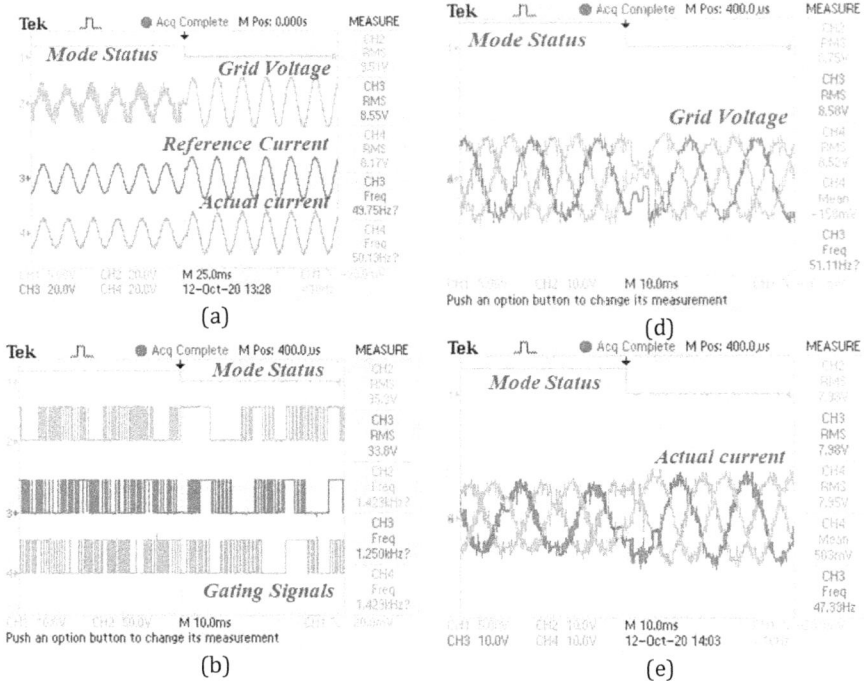

(a)

(b)

(d)

(e)

134

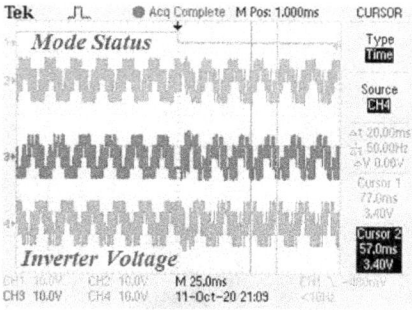

(c)

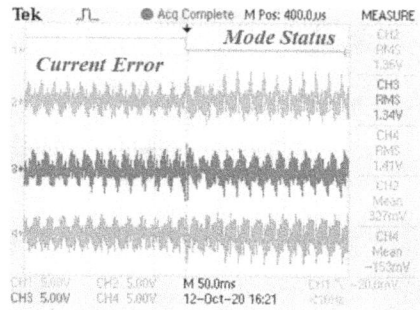

(f)

Figure 4.21: Intra-mode transition within GFC (a). Grid voltage and currents, (b). Gating signals for top three inverter switches (c). inverter leg voltages; (d) voltage at PCC; (e) actual current; (f) current errors.

Case (ii). The second inter-mode transition conducted with the following conditions:

TABLE.4.14 SYSTEM SPECIFICATIONS FOR INTRA- MODE TRANSITION CASE (ii)

| IDM condition | Former mode: Grid -forming with *Genmix 5* | Latter mode: Grid-forming with *Genmix 4* |
|---|---|---|
| Power reference | 7 kW at 0.8pf | 5 kW at 0.9pf |
| Priority inputs | $\rho_L$ is enabled | $\rho_H$ is enabled |
| DA suggested $\lambda$ | 0.4 | 0.2 |

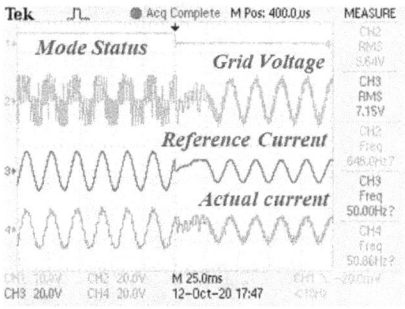

(a)

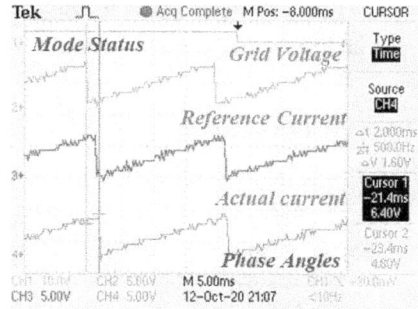

(d)

135

Tek ⎍ ● Acq Complete M Pos: 400.0 µs MEASURE

*Mode Status*

*Gating Signals*

CH2 RMS 35.2V
CH3 RMS 33.8V
CH2 Freq 1.423kHz?
CH3 Freq 1,250kHz?
CH4 Freq 1.423kHz?

CH1 5.00V   CH2 5.00V   M 10.0ms   CH1 20.0mV
Push an option button to change its measurement

(b)

Tek ⎍ ● Acq Complete M Pos: 32.00ms CURSOR

*Mode Status*

*Grid Voltage*

*Reference Current*

*Actual current*

*Phase Angles*

Type **Time**
Source **CH4**
Δt 1.400ms
1/Δt 714.3Hz
ΔV 6.40V
Cursor 1 50.6ms −800mV
**Cursor 2** 49.2ms 5.60V

CH1 5.00V   CH2 5.00V   M 5.00ms
CH3 5.00V   CH4 5.00V   13−Oct−20 16:54

(e)

Tek ⎍ ● Acq Complete M Pos: 1.000ms MEASURE

*Mode Status*

*Inverter Voltage*

CH2 RMS 3.34V
CH3 RMS 3.97V
CH4 RMS 4.34V
CH4 Mean −551mV
CH2 Mean 509mV

CH3 10.0V   CH4 10.0V   M 25.0ms   CH3 25.0mV
12−Oct−20 20:36

(c)

Tek ⎍ ● Acq Complete M Pos: 0.000s MEASURE

*Mode Status*

*Current Error*

CH2 RMS 610mV
CH3 RMS 553mV
CH4 RMS 632mV
CH4 Mean 75.7mV
CH3 Mean 10.0kV?

CH1 5.00V   CH2 2.00V   M 25.0ms
Push an option button to change its measurement

(f)

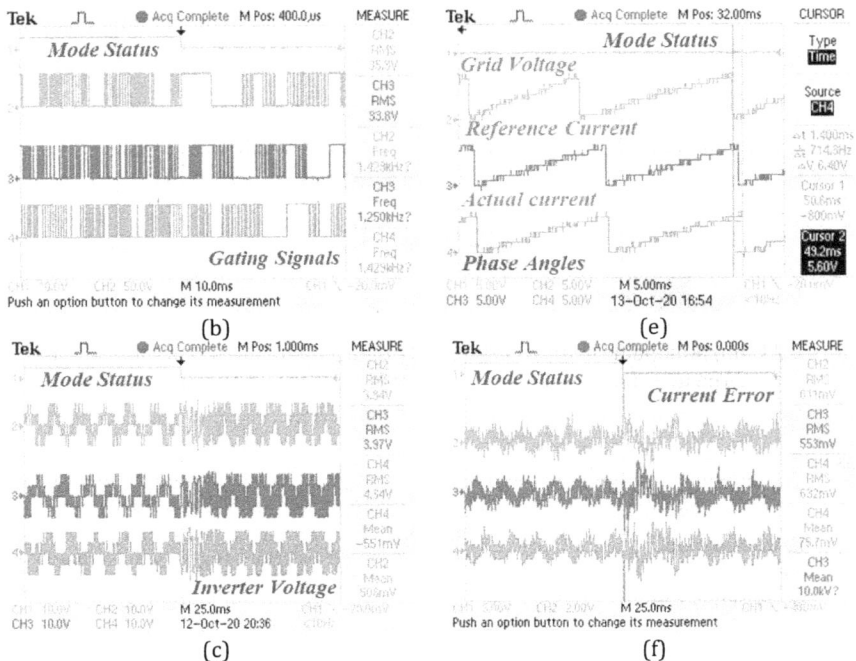

Figure 4.22: Intra-mode transition within GFC (a). Grid voltage and currents, b). Gating signals for top three inverter switches (c). inverter leg voltages; (d) and (e) phase angles (f) current errors.

The DAMPC performance in case ii is similar to that of case i, as depicted in Figures 4.22 (a) to (f). Repeatability is found in all the trends of this cased as discussed in the previous cases.

- **Genmix 4 to Genmix 5**

Case (iii). In case iii, the specifications of the test condition is given as:

TABLE.4.15 SYSTEM SPECIFICATIONS FOR INTRA- MODE TRANSITION CASE (iii)

| IDM condition | Former mode: grid-forming with *Genmix 4* | Latter mode: grid-forming with *Genmix 5* |
|---|---|---|
| **Power reference** | 6 kW at 0.95 pf | 9 kW at upf |
| **Priority inputs** | $\rho_H$ and $\rho_T$ are enabled | $\rho_L$ is enabled |
| **DA suggested λ** | 0.1 | 0.3 |

While DAMPC had to address transition from *Genmix 5* to *Genmix 4* in cases (i) and (ii), with increase or decrease in power transfer, case (iii) represents a transition from *Genmix 4* to *Genmix 5*. Some salient results for case 3 have been depicted in Figure 4.23. (a) to (f).

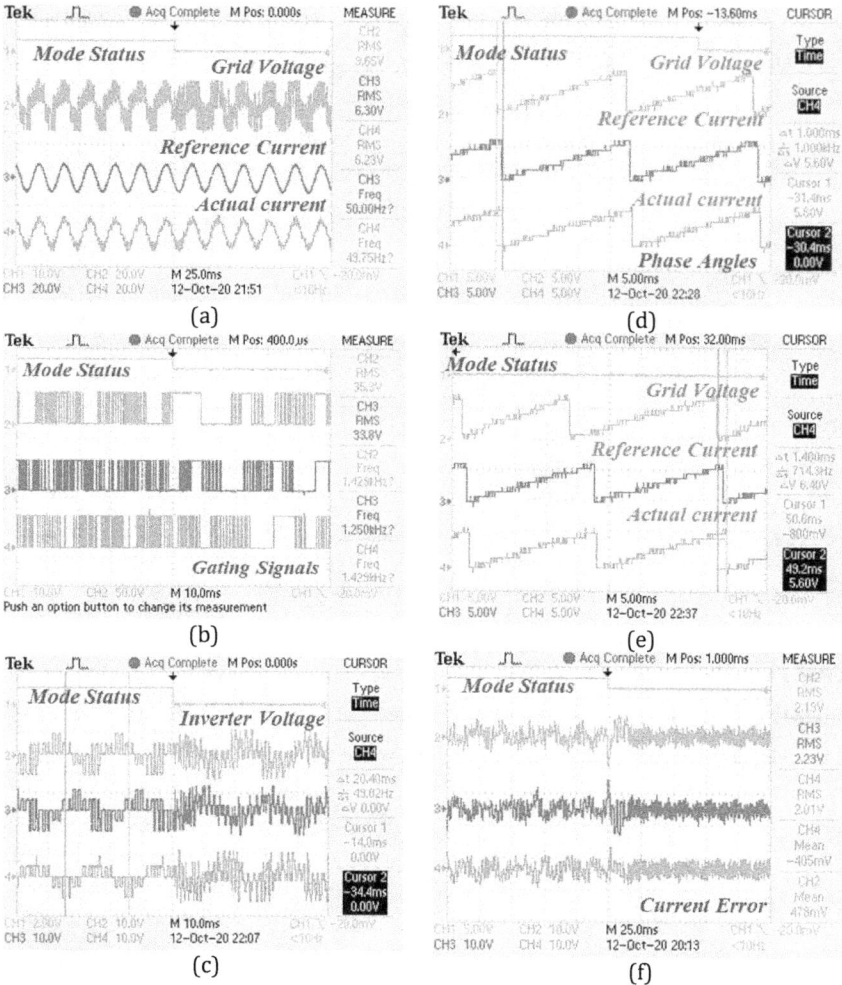

Figure 4.23. Results of intra-mode transition within GFC: (a). Grid voltage and currents; (b). Gating signals for top three inverter switches (c). inverter leg voltages; (d) and (e). phase angles of grid voltage, reference current and actual current for former mode and latter mode; (f) current errors.

As is clear from Figures 4.23 (a) to (f), DAMPC performance in case (iii) is similar to those of cases (i) and (ii) and exhibits similar trend in the tracking error and the choice of λ.

Table 4.16 summarizes all the performance indices corresponding to all the intra-mode transitions within grid-forming mode. The trends of relationship across various parameters are repeated across all the test conditions and all the modes. Across all the cases, the inverter current and voltage *THDs* are found to be within the grid code stipulations. All the performance

137

indices across all the cases indicate a highly competent performance of the developed DAMPC with GTC and GFC. The range of variation of any index across the cases is found narrow and proves the effectiveness of the DAMPC.

TABLE.4.16 HARDWARE TEST RESULTS

| Performance indices | Intra-mode transition: grid-tied to grid-forming | | | | | |
|---|---|---|---|---|---|---|
| | Case (i) | | Case (ii) | | Case (iii) | |
| | Former mode: Grid - forming with Genmix 5 | Latter mode: Grid- forming with Genmix 4 | Former mode: Grid - forming with Genmix 5 | Latter mode: Grid- forming with Genmix 4 | Former mode: Grid - forming with Genmix 4 | Latter mode: Grid- forming with Genmix 5 |
| Power reference | 7 kW at upf | 9 kW at 0.9pf | 7 kW at 0.8pf | 5 kW at 0.9pf | 6 kW at 0.95 | 6 kW at 0.9pf |
| Priority inputs | $\rho_T$ | $\rho_H$ | $\rho_L$ | $\rho_H$ | $\rho_H$ and $\rho_T$ | $\rho_L$ |
| DA suggested $\lambda$ | 0 | 0.2 | 0.4 | 0.2 | 0.1 | 0.3 |
| $F_{asw}$ (kHz) | 1.26 | 1.06 | 0.94 | 1.22 | 1.1 | 0.98 |
| $THD_i$ (%) | 2.01 | 2.14 | 2.87 | 2.02 | 3.02 | 3.9 |
| LOH | 35 | 43 | 23 | 41 | 31 | 37 |
| $\varepsilon_i$ (%) | 0.41 | 2.12 | 0.90 | 0.37 | 0.57 | 0.75 |
| $\varepsilon_v$ (%) | 0.7 | 0.3 | 2.3 | 1.5 | 1.1 | 2.3 |
| $THD_v$ (%) | 2.46 | 1.87 | 4.26 | 2.35 | 3.02 | 4.18 |

## (c) Test results for mode transfer time:

The mode transition times of the DAMPC at sample level and cycle level for both inter and intra-mode transitions are captured and presented in Figure 4.24 (a) to (f).

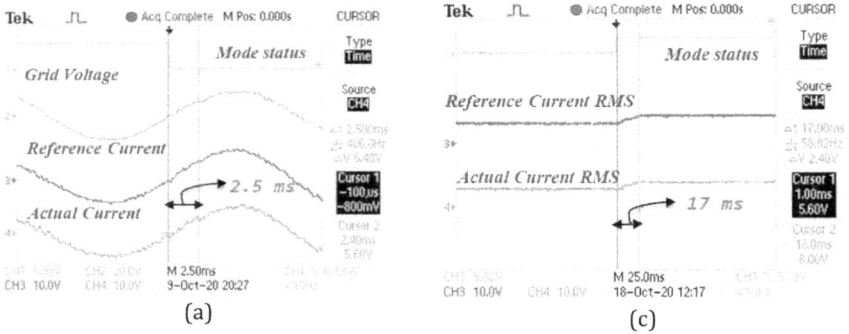

(a)

(c)

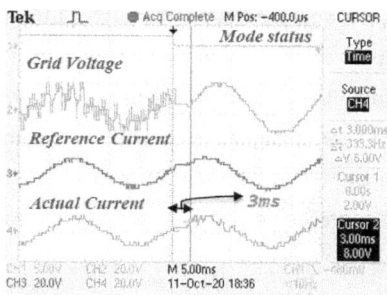

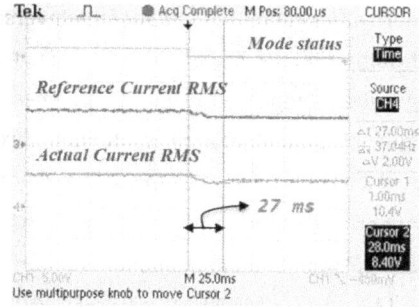

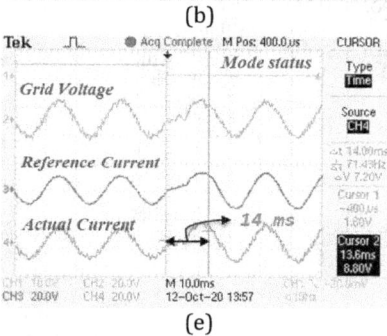

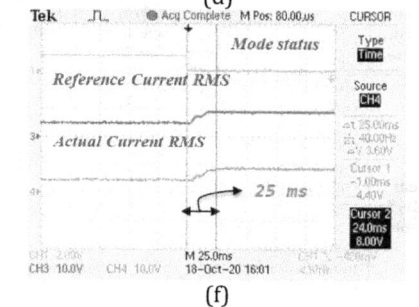

Figure 4.24: Mode transition times at sample and cycle level of DAMPC for (a) to (d) inter-mode transition and (e) and (f) intra-mode transition.

While the three cases of inter-mode transition have the values ranging from 2.5 ms to 14 ms, the intra-mode transition times vary from 17 ms to 27 ms. In intra mode transition, the system impedance at every mode is found to influence the transition time. The modes with higher impedance have resulted in higher transition times and vice versa. Similarly, the percentage step change in the power reference before and after the transitions also found influencing the transition times. Large percentage change in either direction has resulted in large transition times.

## 4.6. Performance comparison of DAMPC with other controls

Finally, a study has been made to compare the performance of DAMPC with that of the state-of-the-art current control techniques including both linear and non-linear controllers. As the dynamic performance for intra-mode transition (within grid-forming) has not been reported in the literature, the comparison is confined to inter-mode operation (grid-forming to/from grid-tied). Also, there are no published literature on MPC controllers tested for dynamic performance in inter or in intra mode transition during microgrid operation. The inter-mode transition time observed for DAMPC is lower than that of the droop-based linear controllers

139

and more or less at par with that of the Inverse plant model feedforward control [145]–[148]. Significantly, DAMPC has a lower average switching frequency with on par levels of THD in voltage when compared to all other schemes including MPC based grid-forming controls. The flexibility of control objectives attained through priority signals and with a dynamically assigned $\lambda$ has been found as the unique feature of DAMPC. A detailed comparison of various performance metrics of grid forming inverters of different control schemes with those of DAMPC is presented in Table 4.17.

TABLE 4.17. PERFORMANCE COMPARISON OF PROPOSED DAMPC WITH OTHER CONTROLLERS

| Type of control scheme | Specific control technique | PWM scheme or $\lambda$ | $F_{asw}$ (kHz) | THD of $u$ (%) | GTC to GFC | GFC to GTC | within GF | Inference |
|---|---|---|---|---|---|---|---|---|
| | | | | | Mode transition time (ms) | | | |
| Linear control schemes | Modified droop control [145] | Space Vector PWM | 10 | Not reported | 200 | Not reported | Not reported | $F_{asw}$ and transition times reported are higher than those of DAMPC |
| | Synchronous reference frame based droop control [146] | Sine PWM | 10 | | 70 | 65 | | |
| | Inverse plant model feedforward control [147] | | 3.3 | | 18 | 16 | | $F_{asw}$ is higher than that of DAMPC. Transition times reported are a little less than that of DAMPC. |
| Non-linear control schemes | Conventional MPC [148] | NA | 5 | 1.48 | Not reported | | | $THD_v$ is lower than that of DAMPC, owing to the use of LCL filter in the former. $F_{asw}$ is higher than that of DAMPC |
| | Improved MPC and Periodic MPC [148] | $\lambda$ is fixed during controller design | | 1 | | | | |
| | MPC scheme with modulator [148] | Space Vector PWM | | 3.5 | | | | Both $F_{asw}$ and $THD_v$ are higher than those of DAMPC |
| | DAMPC | $\lambda$ is dynamically assigned during the operation | 1.1 | 3.01 | 24 | 30.91 | 53.23 | $F_{asw}$ is the lowest amongst all, transition times are close to the lowest reported value and $THD_v$ is comparable (and within permissible limits) |

## 4.7. Summary

This chapter presented the design and development of DAMPC inverter with hybrid operating characteristics which can be customized for any IDM. The design and testing of the developed DAMPC controller scheme have been carried out in simulation as well as in real time hardware and the results are discussed in detail. Besides serving as a proof of concept to this thesis, the results establish the research contributions made through. Salient observations from this chapter has been summarized below:

- The MPC with two dedicated controllers, GTC and GFC have been developed to realize multiple operating modes and the transition between them.

- A multi objective weight function to maximize efficiency of the inverter by reducing switching frequency is formulated and implemented in both grid-tied and grid-forming modes.

- The customized DA database developed through sensitivity analysis on the IDM, when employed in the selection of MPC design parameters, abated loss of tracking under all operating conditions besides ensuring accurate reference tracking.

- A rule base with three priority inputs for harmonic control, inverter switching loss and reference tracking has been developed for dissemination of the DA database. The developed DAMPC inverter exhibited self-reliant operation, thus its hybrid operating characteristics can be customized for typical IDM.

- The dynamic adaptability elevates the inverter control to be self-adaptable for dynamically changing environment in IDM that demand frequent transition between operating modes.

- DAMPC achieved inter and intra mode transitions with narrow mode transition times ranging between 2.5 ms to 14 ms and 17 ms to 53.23 ms respectively.

- The developed controller will be helpful in handling the IDM issues related to low inertia, wide band dynamics, uncertain voltage/frequency profile and varying generation mix.

# Chapter 5

# Conclusion

## 5.1 Summary

Migration of legacy electric utility into power electronic dominated microgrids is the quantum leap contemplated for the future. The important asset to be metamorphosed with judicious control capabilities to suit this futuristic environment is the grid tied inverter. The new age power quality issues like frequency jump, phase jump, even harmonics, supra-harmonics and inter-harmonics etc. makes the grid signals highly non-linear. The state-of-the-art phase locked loops has been found inadequate enough to process such non-stationary, multi-component grid signals to achieve synchronization. The intricate capabilities of various adaptive signal decomposition techniques have been explored in this thesis to design and develop a Variational Mode Decomposition (VMD) Synchronizer. The VMD grid synchronizer exhibited shorter extraction times in the range of 1.4 ms to 2 ms with high accuracy while effectively tracking the fluctuating signatures of the emerging power grid. The errors in magnitude, phase and frequency of the VMD outputs have been observed to be less than 0.7%, for fundamental frequency in particular, and with tolerable limits in case of higher order frequencies. The developed synchronizer has also been found to exhibit a superior dynamic performance under various transient events; and is quantified in terms of time taken to track the transient events, tracking accuracy during and after the transient event and the capability to reject random events in the retrieved outputs, etc. The VMD synchronizer, developed not only stands out with imposing features, but also opens a gateway through which other capable signal processing tools can be further explored for use in this key research area.

The present-day inverters are predominantly grid-tied, which deliver synchronized and controlled active and reactive powers. Apart from independent control of active and reactive power, inverters are anticipated to meet additional control functionalities in inverter dominated microgrid. High efficiency with highest performance, bidirectional power flow, independent grid-forming capability, self-adaptability, etc. are some of these advanced functionalities. Model predictive control has been identified as a multi-objective optimization that can accomplish these multiple control tasks through weighted secondary targets along with the primary target (reference tracking) in a single cost function.

First, an investigation on the influence of sampling rate of model predictive control and its performance on a grid-tied inverter has been conducted. Higher sampling frequencies with closer current predictions resulted in better tracking accuracy. The inverter currents in all test cases, though resulted low THDs, exhibited a wide harmonic spectrum up to half of the sampling frequency. Another important observation from the study is on the average switching frequency which is seen as $1/8^{th}$ of the sampling frequency, which helped to reduce inverter loss. The proficiency in dynamic response demonstrated by MPC in prediction as well as estimation substantiates the preference of this technique in future converter controls.

Then, a bidirectional model predictive control for transition between inverter/converter mode operation of grid-tied inverter has been developed. Here the complex outer voltage loop has been excluded which improved the dynamic response of the control as it does not require any PI controllers and their tuning. The developed bidirectional model predictive control has been observed to possess high tracking accuracy, better harmonic profile with reduced power loss. Furthermore, the control exhibits a superior dynamic response when drifting between the two modes typically in the range of 0.4 ms to 1.53 ms. With increase in sampling rates, the tracking accuracy and harmonic profile can be improved with further reduction in transition time. These characteristics of bidirectional model predictive control will make it as an asset for the future bidirectional converter control.

The grid-forming functionality is an inverter control that decides reliability of inverter dominated microgrids. Hence, a model predictive controlled grid-tied inverter with grid-forming control has been developed and tested. A multi-objective cost function with a weighted secondary target for reduction of average switching frequency has been considered for developing the grid-forming controller.

Finally, design and development of a self-reliant dynamic adaptability imparted model predictive control inverter with hybrid operating characteristics which can be customized for typical inverter dominated microgrids has been presented. Features like bidirectional power control, independent grid forming control with multi-mode and mode transition were imparted to MPC through dynamic adaptability. The dynamic adaptability database also infused the self-adaptability feature to model predictive control. The entire scheme was validated through simulation and hardware. Inter and intra-mode transition times obtained were in the range of 2.5 ms to 14 ms and 17 ms to 53.23 ms respectively.

The highlights of the thesis are:

(i) A signal decomposition technique has been identified as a potential grid synchronizer in replacement of the phase locked loops and proof of concept established.

(ii) An inverter control to address the high dynamics of the Inverter Dominated Microgrids has been developed using Model Predictive control and tested in hardware-in-loop with OP4200.

The synchronizer is designed with Variational Mode Decomposition techniques. The VMD parameters are tailored to accommodate the wideband frequency dynamics, reject any complex harmonic spectral content and develop immunity to power system transients. The inverter control developed has advanced features like, multimode operation and transition control, self-adaptability for the high dynamics of IDM, dynamic parameter assignment through priority inputs and customized database.

The major learning experience in this research can be summarized as:

⅄ Signal Decomposition Technique is effectively used as grid synchronizer for grid-tied converters, which can bring a versatility in converter control thanks to the innate immunity to multiple power quality issues which are foreseen in the future utilities.

⅄ The capabilities of signal decomposition algorithms can be helpful to other allied sectors of power electronics such as harmonic measurements, energy calculations of inverter fed powers etc.

⅄ With multi-mode operation and mode transition the inverters of the future will be interoperable between intentional islanding and optional grid tying and can extend additional support in load management to utility.

⅄ The Dynamic Adaptability can bring forth a flexible tradeoff between the harmonic penalties and the net energy cost for future, 100% renewable energy microgrids.

## 5.2 Research Contributions

The major research contributions of this thesis are:

1. Identified the potential of signal decomposition techniques as a new grid synchronization tool and developed VMD synchronizer for IDM.

2. Newly developed an additional functionality named as Dynamic Adaptability which makes the MPC based grid-tied inverters self-adaptable.

3. Formulated multi-mode control and mode transition with mode specific self-adaptivity in grid-tied inverters.
4. Developed a laboratory facility using OPAL-RT for testing and validation of grid-tied power converter control.

## 5.3 Future work

The need for high-speed processors with large memory can be an obstacle for VMD implementation in real time converter control. So, the VMD Optimization time can be further reduced by identifying a suitable alternative to Weiner filter function and ADMM. This will minimize the high-end computational overhead; this can also result in faster synchronization even with reduced sampling frequency.

VMD can be extended for allied power quality applications like active filters, power computation with non-sinusoidal quantities, active islanding detection, etc.

Online-model-updated DAMPC is yet another future work suggested to track the structural change of IDMs.

DAMPC with additional secondary functions for harmonic spectrum shaping and reactive power compensation can also be a future work.

www.ingramcontent.com/pod-product-compliance
Lightning Source LLC
Chambersburg PA
CBHW071646210326
41597CB00017B/2126